D1554622

ROCK WEATHERING

Monographs in Geoscience

General Editor: Rhodes W. Fairbridge

Department of Geology, Columbia University, New York City

B. B. Zvyagin
Electron-Diffraction Analysis of Clay Mineral Structures—1967

E. I. Parkhomenko
Electrical Properties of Rocks—1967

L. M. Lebedev
Metacolloids in Endogenic Deposits—1967

A. I. Perel'man
The Geochemistry of Epigenesis—1967

S. J. Lefond
Handbook of World Salt Resources—1969

A. D. Danilov
Chemistry of the Ionosphere—1970

G. S. Gorshkov
Volcanism and the Upper Mantle: Investigations in the Kurile Island Arc—1970

E. L. Krinitzsky
Radiography in the Earth Sciences and Soil Mechanics—1970

B. Persons
Laterite—Genesis, Location, Use—1970

D. Carroll
Rock Weathering—1970

In preparation:

A. S. Povarennykh
Crystal Chemical Classification of Minerals

E. I. Parkhomenko
Electrification Phenomena in Rocks

R. E. Wainerdi and E. A. Uken
Modern Methods of Geochemical Analysis

ROCK WEATHERING

Dorothy Carroll
U. S. Geological Survey
Menlo Park, California

With a Foreword by
George V. Keller
Colorado School of Mines
Golden, Colorado

PLENUM PRESS • NEW YORK – LONDON • 1970

Library of Congress Catalog Card Number 77-107534
SBN 306-30434-1

A Division of Plenum Publishing Corporation
227 West 17th Street, New York, N.Y. 10011

United Kingdom edition published by Plenum Press, London
A Division of Plenum Publishing Company, Ltd.
Donington House, 30 Norfolk Street, London W.C.2, England

Printed in the United States of America

For Earth Scientists

"For me to write a book. . .is to have an adventure with my readers, both known and unknown."— *Paul Tournier*

IN MEMORIAM

Dorothy Carroll
1907-1970

With the final page proofs of this book on her desk in the U.S. Geological Survey offices at Menlo Park, California, Dorothy Carroll died after a long illness while still a relatively young person. Always young in spirit, she was a pioneer in several fields. She was born July 6, 1907 and raised in Western Australia, attending the University there, graduating in Geology under the guidance of Professor E. deCourcy Clarke, and continued at Imperial College, London University (England). She was, I believe, the first girl from Western Australia to earn the degree of Doctorate of Philosophy in geology. Entering the relatively new branch of science that dealt with the geochemistry and mineralogy of sediments, she worked first on the soils of her native state. She taught briefly at Bryn Mawr College (1951–52). Later she received an appointment to the U. S. Geological Survey in Washington where she set up a full-scale laboratory for this work. Only in her last years did she move to take charge of the Marine Laboratory at the California branch. She had been a member of the Society of Economic Paleontologists and Mineralogists since 1945, of the Geological Society of America since 1954, the Mineralogical Society, the Geochemistry Society, and other learned groups, where she will be sadly missed.

Rhodes W. Fairbridge

FOREWORD

Soil science is perhaps one of the oldest practical sciences, having been of concern to man probably from the time he progressed from a strictly predatory life to one in which agriculture became important. In view of the antiquity of concern with the subject, it is perhaps surprising that it can be approached from a fresh viewpoint, as is done in this book. Because soil science is an applied science, it is not surprising that the approach is usually descriptive, rather than imaginative. For agriculturalists and other land users, perhaps the most important part of soil science is the description of soils and the capacities of such soils to maintain crops, and this is reflected by the fact that soil science is usually treated in a highly descriptive manner, with soil classification being one of the main efforts. The treatment of the subject from a geological point of view, with considerable emphasis on the evolution of soils and the reasons governing their composition and form, makes this a highly readable book.

Books on soil science are timely, with present-day concern with such major problems as the pollution of our environment and the possibility of overreaching our capacity for producing food for an expanding population. Consideration of the best use of soils in producing foodstuffs will require that more agriculturalists be concerned with soil science not only from the descriptive point of view, which allows them to select the best land for food production, but also from the point of view of soil development, so they can better appreciate how a soil becomes depleted with use, and how a soil which is marginal can be supplemented to be made productive. In the area of pollution engineering, soils in a sense form one half of the universe with which man comes in contact and can contaminate directly, the other half being the atmosphere. Here, soil science should be of concern to many branches of engineering and behavioral science, where it has not been considered previously. This book makes an excellent introduction to the subject for readers in these and other fields.

Golden, Colorado
April, 1970

GEORGE V. KELLER

PREFACE

Part of the material in this book was originally brought together for lectures to agricultural students at the University of Sydney, New South Wales, Australia, and later the original lectures were modified and enlarged for lectures in earth science to geology students at American University, Washington, D.C. The information about soils is probably familiar to soil scientists and to agricultural scientists, but no book on weathering is available that gives a general picture of the alteration of the outermost part of the earth's crust by chemical processes. This book endeavors to bring together the results of research from several disclipines to explain the fundamental geochemical principles of weathering and soil formation. The book can be used as an outline on weathering processes for earth science and geology students and, through its references, for further information about the processes whereby rocks of the lithosphere are changed by their chemical environment on the earth's surface into saprolites and soils. As the principles of weathering are similar for all rocks, descriptions of the processes and results are primarily given for igneous rocks, and it is implied that the results of weathering are conditioned by the mineralogical and physical conditions of the different kinds of rocks of the lithosphere.

Weathering is that part of geochemical processes that results in the mineralogical adjustment of rocks to the chemical climate of the earth's surface with its atmospheric pressure and low temperatures. The first part of this adjustment is brought about by oxidation; in humid climates this is followed by leaching, and the result is chemical winnowing. In dry climates there is no chemical winnowing and rock material accumulates. Chemical weathering results in a residuum of clayey materials and insoluble minerals; physical weathering results in a residuum of largely unaltered rock material.

In temperate climates the results of weathering are a mixture of chemical and physical weathering products and, except in extreme climates, it is difficult to separate physical from chemical weathering processes. The same principles of chemical leaching and mechanical disintegration apply to all rocks, but the product of weathering is conditioned by the variables that occur in a specific area to give the particular steady-state chemical environment. The major factor of such an environment is the climate. The effects of weathering can be recognized as two stages of a complete process: in the first stage, geochemical weathering alters fresh rocks into rotten rocks or saprolites; in the second stage, soil is formed as a result of biological activity and continued alteration of the saprolitized rock. In saprolitization, rocks from within the earth very slowly come to equilibrium with the temperature and pressure conditions on the surface. At the lithosphere–atmosphere interface oxidizing conditions prevail, and the surface receives water from rain or snow, and radiant heat from the sun. All these processes cause mineralogical and chemical changes in rocks. The results vary in intensity with climate and with the geological time during which the steady-state chemical environment has been in effect. In the tropics, a fresh rock is converted fairly rapidly (geologically) into a saprolite, but the degree of saprolitization decreases for an equal time interval with change of climate toward the poles, and with altitude. Saprolitization also naturally decreases as altered rocks are removed by erosion. In dry climates, where water seldom or intermittently leaches rocks, saprolitization does not occur, although fossil saprolites, indicating earlier more humid conditions, may be present. In some desert climates rainwater and groundwater never penetrate rocks, and there weathering consists of oxidation and the physical disintegration of minerals and rocks.

This book attempts to give a broad outline of the weathering processes and their results. It presupposes that its readers have some knowledge of rocks, minerals, chemistry, and geochemistry. Chapter 1 discusses the composition of the earth's crust (lithosphere) and the equilibrium conditions of the surface. This is followed (Chapter 2) by a description of the weathering environment in terms of pH and Eh, of climate and microclimate, and of the five variables that have been recognized in describing the weathering steady state. Chapter 3 describes geochemical and pedochemical weathering. The former produces saprolites, and the latter soils on the surface of saprolites. In Chapter 4 soils and the soil pattern produced by the growth of plants in the saprolite surface, together with the relation of minerals to chemical weathering, are described more fully; and in Chapter 5 present-day soils are described as well as a classification of these soils, and how this classification applies to the soils of the coterminous United States. Chapter 6 summarizes the principal methods used to calculate the amount of rock that has been weathered to form a soil. Chapter 7 discusses the physical (mechanical) disintegration of rocks and gives examples of the sizes of the fragments that have been found at the first stage of the breakup of rocks. Chapter 8 describes chemical weathering, the relation of silica and alumina to the weathering environment in processes such as those that produce laterites and bauxites, and it also gives information about other weathering products such as loess, sands, and alluvial soils. Biological activity in weathering and soil formation is treated in Chapter 9, together with some of the organic chemical reactions (chelation) that occur. Information about the temperature in weathering is given in Chapter 10, and in Chapter 11 some of the features brought about by the amount of time involved in weathering are described together with fossil soils and the maturity of weathering produced in any region. The presence and role of minor or trace elements in soils, their relation to the trace element content of rocks, and their importance in plant, animal, and human nutrition are discussed in Chapter 12. Although all weathering is important to the human population of the earth, trace element presence and distribution form one of the most fascinating and interesting aspects of rock weathering. For convenience of reference, Appendix 1 gives the principal mineralogical changes in weathering. Appendix 2 gives the raw data references to investigations that have been made of the change of various rocks into soils. The data have been classified climatically according to Fig. 3 (Section 2-2).

The literature on rock weathering and soil formation is voluminous. Some investigations have been made by geologists, but most of the published information is scattered through soil science publications. A selection of the available source material was made for this book over the course of many years of interest and investigations in a field spanned by geology, soil science, chemistry, and plant and animal nutrition. I am indebted to many scientists for their help and encouragement in my chosen studies of rock weathering and soil formation, but it is not possible to name them all. However, special thanks go to the late Edward de Courcy Clarke, Professor of Geology at the University of Western Australia, for his interest and encouragement when I was a graduate student, to Professor L. J. H. Teakle, University of Queensland, Australia, who taught me much about soils in the field, to Roy Brewer, Soils Division, Commonwealth Scientific and Industrial Research Organization, Canberra, who took the color photographs of Plate 2, to Professor Robert M. Garrels, for many discussions of the chemical weathering of minerals, and to the many students both in Australia and the United States whose need gave me the incentive to prepare the original lectures.

Menlo Park, California
June 3, 1969

CONTENTS

Chapter 1. The Earth's Crust 1

1–1. Rocks and Minerals 1
1–2. Disequilibrium and Equilibrium 5
1–3. Alteration of Original Rocks by Weathering 5

Chapter 2. The Weathering Environment 7

2–1. Fundamental Processes 7
2–2. Climate and Microclimate 9
2–3. The Five Variables of Weathering 15
2–4. The Steady State in Weathering 16

Chapter 3. Geochemical and Pedochemical Weathering 19

3–1. Saprolites 19
3–2. Soils 21

Chapter 4. Soil 25

4–1. Soil as an Entity 25
4–2. Soil Profiles 26
4–3. Mineral Products of Weathering 30
4–4. Residual Minerals of Weathering 33
4–5. Clay Minerals 35

Chapter 5. Soil Patterns 39

5–1. Present-Day Soils 39
5–2. Classification of Soils 41
5–3. Soils of the United States 46
5–4. Lateritic Soils, Laterite, and Bauxite 53
5–5. Volcanic Ash 59
5–6. Loess 61
5–7. Sands 65
5–8. Alluvial Soils 67

Chapter 6. Amount of Chemical Weathering 69

6–1. Calculating the Amount of Chemical Weathering 69
6–2. Barth's Calculations 69
6–3. Calculations from Chemical Analyses 70
6–4. Calculations from Mineralogical Analyses 72
6–5. Calculating the Amount of Clay Developed 75

Chapter 7. Physical Weathering 81

7–1. Mechanical Disintegration 81
7–2. Primary Breakdown into Grains 81
7–3. Effect of Particle Size on Weathering Processes 85
7–4. Grain Sizes in Soils 87

Chapter 8. Chemical Weathering 89

8–1. Weathering by Water 89
8–2. Kinds of Water in Weathering 93
8–3. Solution 97
8–4. Hydrolysis 103
8–5. Ion Exchange 107
8–6. Oxidation and Reduction 112

Chapter 9. Biological Acivity in Weathering 117

9–1. Role of Plants and Animals 117
9–2. Ecology and Climatic Pattern 121
9–3. Organic Matter and the Formation of Humus 122
9–4. Humus in Soil Formation 123
9–5. Complexing and Chelation (Chevulation) 124
9–6. Organic Weathering 127

Chapter 10. Temperature in Weathering 129

10–1. Weathering Temperatures 130

Chapter 11. Time in Weathering 135

11–1. Duration of Weathering Period 135
11–2. Pedological Indication of Time 139
11–3. Fossil Soils 140

Chapter 12. Trace Elements in Weathering 145

12–1. Amounts of Trace Elements in Rocks in the Lithosphere . . 145
12–2. Association with Minerals 145
12–3. Distribution in Granitic and Basaltic Rocks 147
12–4. Chemical Characteristics 147
12–5. Behavior of Trace Elements in Weathering 148
12–6. Petrographic and Biogeochemical Provinces 150
12–7. Relationship of Trace Elements to Plant and Animal Nutrition 154
12–8. Relationship of Trace Elements to Human Nutrition . . . 166

Appendix 1. Mineral Transformations in Weathering 169

Appendix 2. Data of Rock Weathering 175

Literature Cited 189

Index 201

Chapter 1

THE EARTH'S CRUST

The part of the earth on which we live is the weathered crust of the lithosphere, the surface of our planet. Beneath the lithosphere are the mantle and the core of the earth. The crust is the exposed surface of the lithosphere that has been modified in various ways by climate and by the vegetation that forms a cover where the climate is favorable to the growth of plants. According to the climate we find regions of tundra, prairies, semitropical and tropical forests, and deserts. About 70% of the lithosphere is covered by water, but the remaining 30% is dry land in which we are interested. The part of the earth that we see is the uppermost surface of the dry land; it is a layer of material of variable thickness that has been produced from the hard lithosphere by a combination of chemical and physical processes acting on it with different intensity at low temperatures (up to about 30° C) for varying lengths of time. The results of these chemical and physical changes are known as weathering. The expression of the changes caused by weathering vary with the kind of rock, the kind of climate, and the time that the local conditions have prevailed. Generally the effects of weathering are shown by the presence of altered rocks and minerals, saprolites (rotten rocks), and soils.

1–1. Rocks and Minerals

It has been calculated that the upper 10 miles of the lithosphere consists of 95% igneous rock (about equal amounts of granite and basalt), 4% shale, 0.75% sandstone, and 0.25% limestone. The composition of this part of the lithosphere was calculated by Clarke and Washington (1924) from over 5000 analyses of fresh rocks to be: SiO_2, 60.18%; Al_2O_3, 15.61%; Fe_2O_3, 3.14%; FeO, 3.88%; MgO, 3.56%; CaO, 5.17%; Na_2O, 3.01%; K_2O, 3.19%; TiO_2, 1.06%; P_2O_5, 0.30% (average of analyses recalculated to 100% without water or minor constituents). The principal part of the lithosphere consists of igneous rocks that have consolidated from molten rock material known as magma. The temperature and pressure within the lithosphere keep magma in a viscous state. Magma is classified according to its chemical composition as siallic (Si, Al) and mafic (Mg, Fe). Siallic magma, with over 50% silica, consolidates into granitic rock; mafic magma, with less than 50% silica, consolidates into basaltic rock. Granitic rocks are of deep-seated origin within the lithosphere and their presence on the earth's surface is caused by erosion. Basaltic rocks, however, form from lavas that spread from volcanoes on the earth's surface. Examples of such lavas are the basaltic rocks of the states of Washington and Oregon, and of the Hawaiian Islands.

When magma cools it solidifies and its chemical constituents crystallize as minerals. Granitic rocks contain an abundance of quartz and feldspar and are light in color, but basaltic rocks are predominantly dark-colored.

A useful field classification of igneous rocks is that given by Moorhouse (1959, p. 151), in which these rocks are grouped according to their content of dark minerals (Table 1).

Rocks are quite complex both in their mineralogy and texture or fabric. The detailed study of rocks is the field of petrography, but we need to know the main characteristics of rocks in order to learn something about the geochemistry of rock alteration. Each mineral in a rock has a definite chemical composition and atomic structure which causes its crystalline form and physical properties such as hardness, resistance to alteration, specific gravity, cleavage, etc. Minerals are packed together in rocks, but each mineral fragment exhibits its own optical and chemical characteristics.

Eight chemical elements account for nearly all the principal minerals. These elements are oxygen (46%), silicon (28%), aluminum (8%), iron(5%) magnesium (2%), calcium (3.5%), sodium (3%), and potassium (2.5%). The remaining elements in the periodic table are minor so far as the overall composition of the lithosphere is concerned.

Inasmuch as oxygen accounts for nearly half of the chemical composition of the lithosphere, it is interesting to note that the percent by volume of oxygen in a number of materials is: air, 20.95%; ocean, 96.87%; quartz, 98.73%; average igneous rock, 91.83%; average basalt, 91.11%. The space taken up by most minerals is due to oxygen atoms. In most rocks oxygen makes up 92% by volume, and all cations present (Si, Al, Fe, and others) account for the remaining 8%.

The principal minerals in igneous rocks are *feldspars*, complex silicates of aluminum with potassium (orthoclase and microcline), or sodium and calcium (the plagioclases, albite to anorthite), quartz (SiO_2), ferromagnesians (silicates of calcium, magnesium, and iron), the micas, with complex frameworks of alumina and silica to which are bonded potassium (muscovite) or magnesium (biotite and vermiculite), the amphiboles, and the pyroxenes.

TABLE 1

Content of mafic minerals	Quartz over 5% of felsic minerals	Quartz less than 5% of felsic minerals
0–20%	Granitic rocks and varieties	Syenitic rocks and varieties
20–40%	Quartz diorite and varieties	Diorite and varieties
40–75%	Quartz gabbro Quartz diabase Quartz basalt	Gabbro Diabase Basalt
Over 75%	Pyroxenite Hornblendite Peridotite	

Felsic minerals—light-colored, feldspars.
Mafic minerals—dark, ferromagnesian minerals.

In addition to the essential minerals of rocks there are *accessory* minerals in small quantities and generally in small grains. Among these accessory minerals, many of which are characteristic of certain kinds of rock, are ilmenite ($FeO \cdot TiO_2$), magnetite ($FeO \cdot Fe_2O_3$), apatites (a complex group of minerals that are essentially phosphates of calcium with chlorine and fluorine), zircon ($ZrSiO_4$), sphene ($CaTiO_3$), and others. Many of these accessory minerals are very resistant to weathering, and hence appear practically unaltered in weathering products.

The mineralogy of igneous rocks is complex, and numerous different species of minerals are found in the various kinds of rocks. The principal rock-forming minerals associated with granitic and basaltic rocks are as follows:

Granitic rocks

Quartz, orthoclase feldspars, micas (muscovite and biotite), ±hornblende

Basaltic rocks

Plagioclase feldspars, pyroxenes, ± olivine

The chemical elements in the two major classes of rocks are arranged in this way:

SiO_2, about half present as quartz, the remainder in other minerals; Fe, very little, in micas and hornblende; Al_2O_3, in feldspars and micas; TiO_2, a little, in micas

All the SiO_2 is combined in feldspars, pyroxenes, olivine; Fe, much, in pyroxenes, sometimes in olivine, magnetite, and ilmenite; Al_2O_3, in feldspars and pyroxenes (there is more Al_2O_3 in basaltic rocks than in granitic rocks); TiO_2, often much, in pyroxenes and ilmenite

The occurrence of the major chemical elements in the common minerals of these kinds of rocks is shown diagrammatically in Fig. 1.

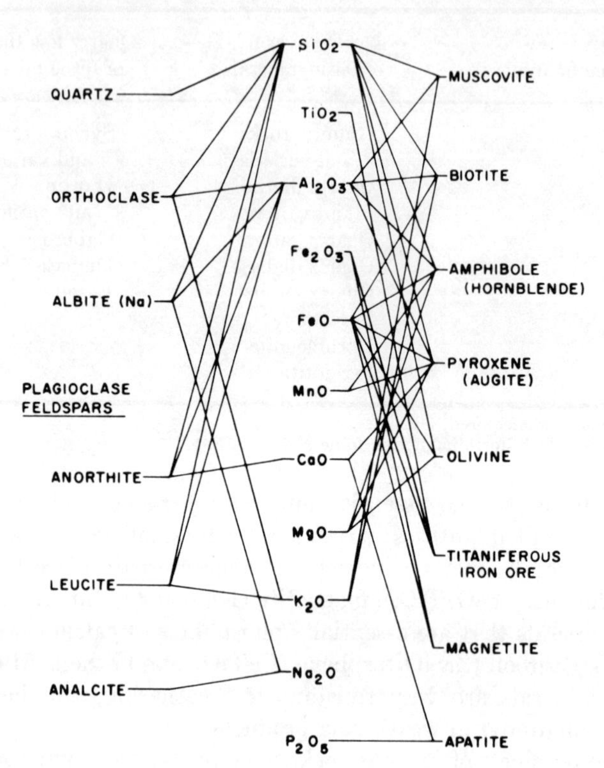

Fig. 1. Association of chemical elements with common minerals (after Mohr, 1944).

In addition to igneous rocks metamorphic and sedimentary rocks are of importance in rock weathering as they are exposed on the earth's surface in many places.

The metamorphic rocks contain, in addition to feldspar and quartz, such minerals as amphiboles, chlorites, micas, garnet, andalusite, staurolite, kyanite, and sillimanite, which have crystallized in the rocks under the conditions of pressure and temperature induced by the metamorphism. For example, there are chlorite–garnet schists, actinolite schists, and kyanite–sillimanite schists. The minerals produced by metamorphism are generally very resistant to weathering.

The sedimentary rocks contain as detrital grains such minerals as quartz, mica (of the variety generally called illite), and minerals that have formed *in situ* like calcite. In addition they contain unweathered grains of accessory minerals of igneous rocks, and grains of aluminosilicates produced in metamorphic rocks.

The *provenance* (origin) or derivation of the detrital minerals of sedimentary rocks is difficult to unravel. The composition of the detritus depends on the distributive province that was eroded to form the sediment. For example, the reworked Atlantic Shelf (North America) marine sediments contain minerals that were removed from New England by glaciation as well as those of eroded and transported materials of Cretaceous beds of the northern Atlantic Coastal Plain and hinterland.

1–2. Disequilibrium and Equilibrium

The chemical constituents of igneous rock magma are in equilibrium with the conditions of temperature and pressure under which the magma forms. The chemical elements may crystallize within the magma, but if the conditions change minerals form whether the magma is granitic or basaltic. As a magma cools disequilibrium occurs, and the minerals formed either at the time of cooling or earlier are in a metastable condition. Alteration by weathering is the expression of the minerals of an igneous rock adjusting to the equilibrium of air and water at the earth's surface. The adjustment is generally extremely slow and has to be measured by geological time scales. Emplacement of igneous rocks may be very slow, as in granitic rocks which may be beneath considerable thicknesses of other rocks, or it may be rapid and on the surface, as in the lava from volcanoes, the lava on the island of Hawaii being an example. The change of state of the minerals in a magma is slow in a granitic rock but rapid in basaltic rocks. The minerals of both kinds of rocks can be considered metastable. The durability of granite is well known, and one is apt to think of solid rocks as durable. However, the fact that such rocks on the surface of the earth are under different conditions of temperature and pressure than those under which they formed shows that they are in disequilibrium. The unroofing or uncovering of an igneous body like granite is nearly always shown by the development of joints which may be contractual, produced when the rock cools, or expansional, as when a rock mass is relieved by erosion of the weight of overlying rock. Lava flows that solidify as basalt tend to develop columnar jointing as they cool, the most famous example of which is the Giants' Causeway in County Antrim in Northern Ireland. Joints may not be conspicuous until a rock begins to be altered by its environment, but joints cause weathering (adjustment to equilibrium) to begin by forming a path for the movement of water which initiates chemical weathering.

1–3. Alteration of Original Rocks by Weathering

Rocks are altered by contact with the atmospheric conditions at the earth's surface. The alteration is fundamentally a chemical process in which the original silicate minerals are affected by inorganic and organic solutions low temperatures. The reactions that take place between silicate minerals and solutions are governed by the laws of chemical solution and equilibrium, free energy, and redox potentials. The results of the sum total of the chemical reactions are governed by a number of variables that result in a steady-state chemical environment in any particular region of the earth's surface.

The effect of the various agents and processes of weathering reacting with rocks is shown by mineralogical, chemical, and grain-size changes in the weathered material if it is compared with the unweathered rock. The changes produced in the fresh rock by weathering can be ascribed to partial or complete decomposition of some minerals; the stability of other minerals; the oxidation of ferrous (Fe^{+2}) to ferric (Fe^{+3}) iron; and to the partial or complete mobilization of both major and minor chemical elements. In part the reactions are caused by the solubility of the constituent minerals, and in part by the porosity of the rock which either augments or retards leaching by water. Saturation of the rock by groundwater is important, as under saturated conditions the minerals are soaked or bathed in aqueous solutions, thereby promoting conditions and reactions that result in the removal of soluble materials (calcite being an example), reactions between fresh mineral and solution, and reactions between weathered material (such as clays) and minerals. The principal rock-forming minerals vary in their reactions in pure water. Such reactions are generally shown by the pH of the solution. Dilute acids and dilute alkalies react differently with different minerals, and there is a range in the conditions under which a mineral reacts.

In most natural situations the water that is available to react with rocks and minerals has pH values between pH 3 and pH 9, corresponding to a solution containing 1 meq H^+ to one containing 0.000,001 meq H^+ per liter; or a solution that is 0.001 N to 0.000,000,001 N for H^+, or slightly acid to moderately alkaline. However, there is in moist climates a preponderance of situations in which the water is slightly acid to neutral in reaction, so that most chemical weathering occurs under the acid to neutral conditions.

In the first stages of chemical weathering of rocks an alkaline solution is produced by the alteration of feldspar. Most later chemical reactions are acid; silica is gradually leached from the rock leaving an enrichment in alumina. The end product of silica removal is an impure aluminous and often ferruginous material that is known as laterite. If little iron is present in the original rock, then bauxite (aluminous oxides, largely gibbsite) results. Various stages of desilication can be recognized as weathering proceeds. Desilication may proceed at various rates in different parts of a given area because of differences in drainage.

Chapter 2

THE WEATHERING ENVIRONMENT

The alteration of rocks at the earth's surface is shown by the development of saprolites and soils. Rocks vary in composition and therefore in their susceptibility to alteration by weathering processes. Furthermore, the chemical environment in which rocks are altered varies with position on the earth's surface and also with local conditions. The environments under which most atmospheric weathering takes place are oxidizing with an oxidation potential (Eh) of + 600 mV and a pH ranging from pH 4 to pH 10. Environments isolated from the atmosphere may have Eh ranging from + 400 mV to − 400 mV within the same pH range as above. Many chemical reactions between minerals and water take place in the contact of groundwater, water in rocks, waterlogged rocks or soils, and in organic-rich fresh and saline waters. Baas-Becking et al. (1960) compiled the results of many hundreds of published analyses of the pH and Eh values of soils, shallow groundwater, and fresh water to obtain the natural weathering environment as it actually occurs. Figure 2 shows this environment.

2–1. Fundamental Processes

The atmosphere, biosphere, and hydrosphere are the reactors in a vast chemical process with the surface of the lithosphere to produce the earth's crust. Each reactor consists of a complex chemical process that is intimately associated with physical processes. For the chemical reaction we may write:

$$\text{(Atmosphere + biosphere + hydrosphere) + lithosphere} \rightarrow \text{weathered lithosphere + residual materials + dissolved chemical elements}$$

The physical breakdown of rocks accompanies the chemical process of weathering, and its intensity varies according to the climatic situation (Section 7–1). The chemical elements that are released by mineral alteration are largely removed by chemical weathering.

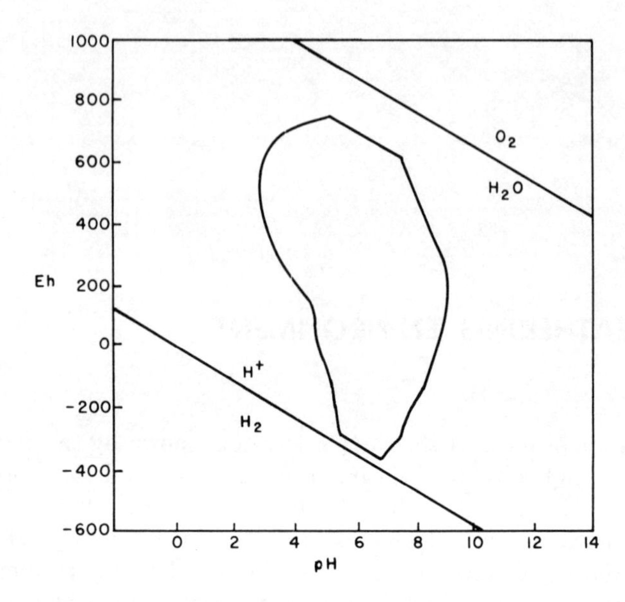

Fig. 2. Weathering environment of soils, shallow groundwater, and fresh water as characterized by pH and Eh (after Baas-Becking et al., 1960).

The fundamental process of the weathering environment is leaching by water, largely supplied by rain; contact with the atmosphere provides oxidation. Both processes are discussed in detail in Sections 8–1 and 8–6. It is significant that, apart from oxygen, the two most abundant chemical elements of the lithosphere are silicon and aluminum, and the most abundant minerals, with the exception of quartz, are aluminosilicates of various kinds, so that chemical weathering is principally concerned with aluminosilicate chemistry. The principal products of weathering are also the aluminosilicate minerals that form clays. Silica and alumina differ in their reaction with water: silica is sparingly soluble at the pH range of all natural environments, but alumina is only soluble in acid or alkaline environments, and not at near neutral pH values (see Fig. 23, Section 8–3). Silica in quartz is only soluble to the extent of 6 ppm, and it therefore accumulates, especially in the weathered products of granitic and highly siliceous rocks such as sandstones and siliceous metamorphic rocks that contain quartz.

To relate weathering to the climatic conditions and vegetation of any area a leaching factor has been recognized (Prescott, 1949). Leaching is related to the amount of rainfall and to the amount of water lost by evaporation and transpiration from the vegetation of the area, that is, to the amount of water that is available for weathering. The leaching factor has been studied as a simple ratio, P/sd, where P is the precipitation (rainfall), and sd is the saturation deficit, or relative humidity (evaporation of an open surface of water depends on the relative humidity of the atmosphere; compare evaporation in desert and humid climates). This relation is not simple, and like most weathering processes depends on the interaction of a number of variables; however, steady-state conditions do occur, and data for the various climatic types of weathering can be grouped. From a number of different water-use figures Prescott found that the most efficient single-value climatic index, as shown by the leaching factor, is P/E^m, where E is evaporation and m is a constant with a mean of 0.73. A value for the climatic index of 1.1 to 1.5 corresponds to a point where rainfall balances transpiration from vegetation and evaporation from the weathered crust. If the rainfall is greater, water is available for continued chemical leaching. The climatic index ignores the circulation of water and dissolved chemical elements by vegetation; such movement of materials constitutes an important part of the weathering process (Sections 9–1 and 12–6).

2–2. Climate and Microclimate

The *climate* of any region is the sum of all the weather that occurs there. Weather is the state of the atmosphere at any given time and place. The earth rotates around the sun and tilts on its axis, thereby producing a climatic pattern with a seasonal variation which we recognize as spring, summer, autumn, and winter. There is a definite relation between the position of the earth's tilt and the season, which we see as the position of the sun and the length of the day. The extremes of climate are more easily seen than the changes in the temperate types of climate; thus Arctic and Antarctic climates show marked extremes between summer and winter, but tropical climates are more equable. Climate influences in a major way the types of weathering that will take place in any region. In polar regions the principal weathering is physical, causing the rocks to break into splinters, but in a warm moist tropical climate chemical weathering is at a maximum. The climates of the world can be classified into a number of major groups. One such classification was made by Köppen, from whose map the climatic features of any region can be generalized (Fig. 3).

In Fig. 3 the major climates are designated by capital letters, and variation within each group is designated by a small letter. The data of rock weathering given in Appendix 2 have been grouped according to the climate following Köppen's map.

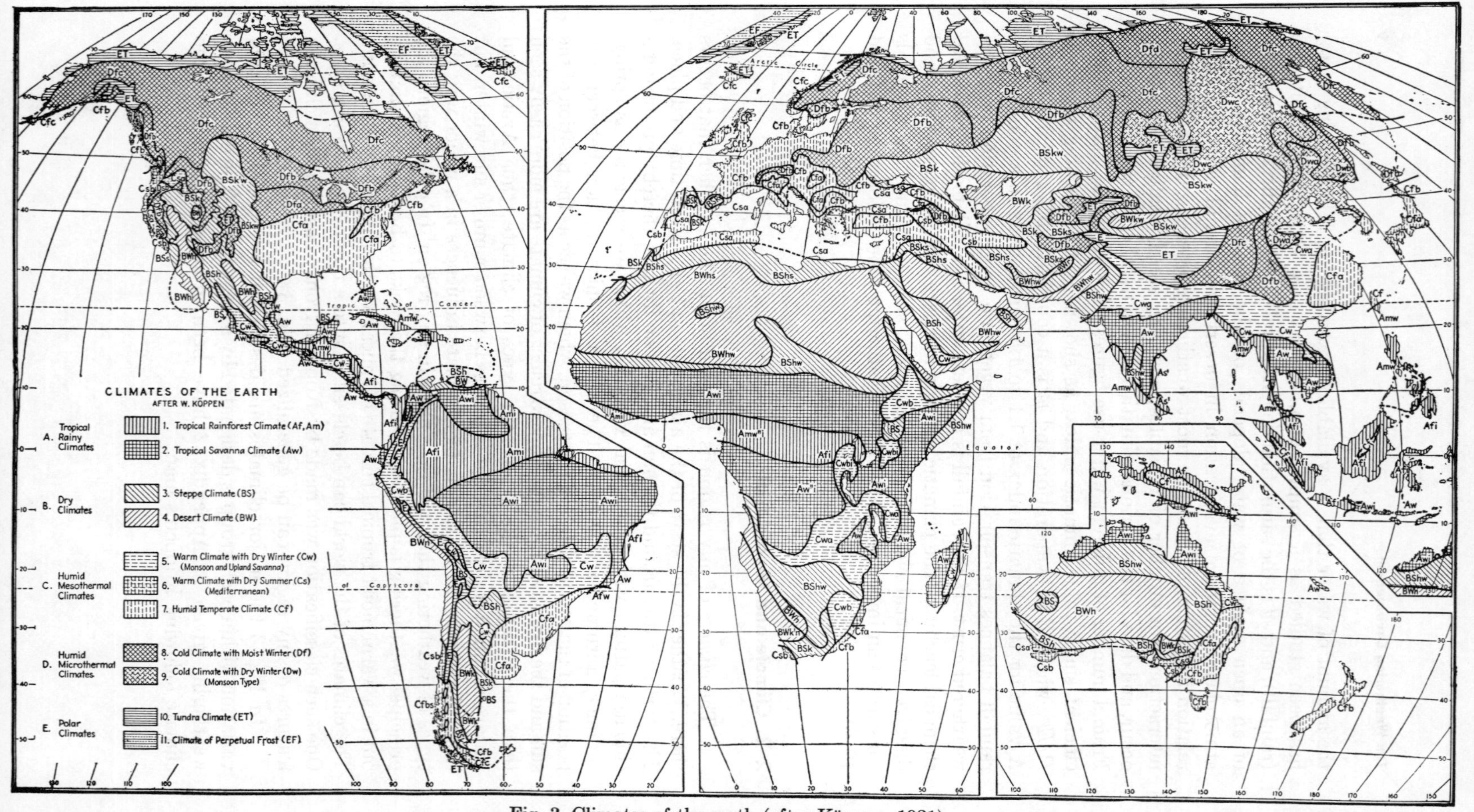

Fig. 3. Climates of the earth (after Köppen, 1931).

The coded symbols of the climates in Fig. 3 have the following meanings:

First letter

- A, C, D, sufficient heat and precipitation for high-growth trees.
- A, Tropical climates, as at Singapore, Malaysia.
- B, Dry climates, as at Aswan, Egypt.
- C, Humid, mesothermal climates: coldest month between 64° and 27°F; Mediterranean type of climate as at Palmero, Italy.
- D, Humid, microthermal climates: warmest month over 50°F, coldest month below 27°F, as at Moscow, USSR.
- E, Polar climates: warmest month below 50°F, as at Nova Zembla.

Second letter

- S, Steppe climate.
- W, Desert climate (German *Wüste*).
- f, Sufficient precipitation each month.
- m, Forest, tropical climate despite dry period.
- s, Dry period in summer of respective hemisphere.
- w, Dry period in winter of respective hemisphere.

Third letter

- a, Warmest month over 72°F.
- b, Warmest month below 72°F; at least 4 months over 50°F.
- c, Less than 4 months over 50°F.
- d, Less than 4 months over 50°F; coldest month below 36°F.
- h, Dry-hot; mean annual temperature over 64°F.
- k, Dry-cold; mean annual temperature below 64°F.

Some examples of places having these types of climate are: San Francisco, Calif., Csb; Denver, Colo., Bsk; Chicago, Ill., Dfa; Washington, D.D., Cfb; Louisville, Ky., and Texas, Cfa; Egyptian desert, Arabia, central Australia, Bwh; Amazon Valley, Am; Greenland, Baffinland, E; Allahabad, India, Cw.

An outline of the main chemical processes that occur in the weathering products of the various climatic zones is given in Table 2.

These weathering characteristics result in the formation of the soils that are considered in more detail in Chapters 3 and 4.

A *microclimate* is the climate near the ground, and is the variation of the major climate of a region; it is the natural environment of plants and animals. Microclimates occur within major climatic zones. One example of a microclimate is a fog belt or pocket such as occurs in the San Francisco Bay area. Man-made microclimates occur in the major cities of the world and, although known to the inhabitants of the cities, are best appreciated from

TABLE 2
Weathering Characteristics of the Major Climatic Zones (Partly after Jacks, 1934)

Soil type	Soil characteristics	Climate	Vegetation	Clay minerals	Exchangeable cations	
Tundra	Peat; subsoil permanently frozen	Perpetually cold	Sphagnum moss	Detritals; mica, chlorite	Probably H > Ca > Mg > Na	Leaching of bases and sesquioxides (chelation important)
Podzol	Acid raw humus; topsoil leached of bases and sesquioxides	Cold moist winter; mild summer; rain evenly distributed	Coniferous forest or heath	Same	H > Ca, Mg > Na	
Brown earth; transitional type	"Mild" humus or mull; some leaching of bases; surface weakly acid	Moist temperate	Deciduous forest	Mica, chlorite	$Ca \geq H$	
Chernozem and other black earths	No raw humus, but deep humus horizon; surface neutral, no leaching; calcareous layer	Continental; cold winter, hot summer, 10–20 inches rainfall; or tropical with restricted drainage	Perennial grasses	Mica and/or montmorillonite	Ca > H	
Chestnut soil	Less humus than in chernozem; weakly alkaline; some accumulation of salts near surface	Continental with low rainfall	Dry steppe; xerophytic shrubs	Mica, chlorite	Ca > H	

Temperature	Rate of humus decomposition	Humidity	Acidity	Leaching		
Saline and alkali soils	Strongly alkaline, and/or accumulated salt in upper horizons; low humus	Arid	Halophytes and thallophytes	Mica, chlorite, attapulgite, montmorillonite	Na > Ca > H	
Desert Latitude 30°N and S	Disintegrated parent material with or without movement by wind and flash floods	Arid	Nil to sparse	Detritals and attapulgite	No leaching	
Tropical red earths	Rather acid; red or reddish; very low humus; some concentrated sesquioxides at surface	Hot; alternate wet and dry seasons	Savannah and tropical rain forest	Kaolinite; mica, chlorite in less strongly leached places	H > Ca > Mg > Na	(Chelation probably important) Leaching of bases and SiO_2
Laterites and lateritic soils	Acid; brick red; practically no humus; leached of bases and much SiO_2; surface consists largely of hydrated sesquioxides	Hot, high rainfall	Tropical forest	Kaolinite, halloysite, gibbsite	H > Ca > Mg > Na	

► = Direction of increase.

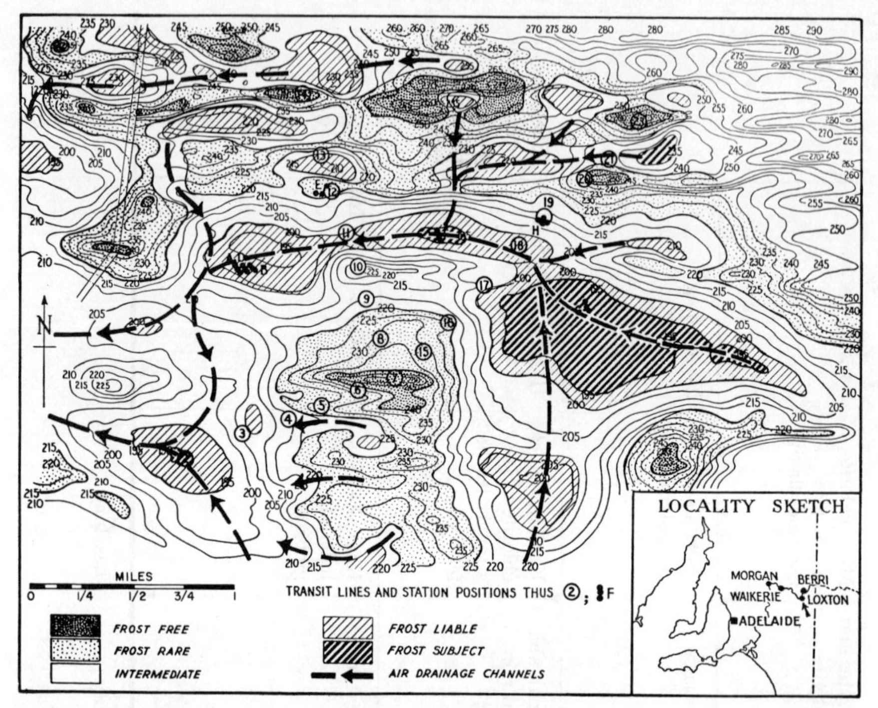

Fig. 4. Microclimate in a small area near Adelaide, South Australia (after Mason, 1958).

the air. Such microclimates have very little influence on the characteristics of weathering of the earth's crust because human populations have congregated after the lithosphere has weathered.

Very small vertical and horizontal changes in the configuration of the earth's surface cause microclimates that are characterized by vegetation and animal changes. Hence there are, under natural conditions, stands of certain kinds of trees and their associated communities of small trees, shrubs, grasses, and herbs, and animals that inhabit an environment favorable to them. Examples are the redwood forests of coastal, northern California, and the live oak–white oak associations in drier inland areas of the same region. Ecological associations are an expression of microclimates, and the study of habitat associations is known as ecology. Microclimates show that although the macroclimate determines the main character of the weathering in any region, the microclimate influences the soil pattern. That is, it expresses the weathering taking place in any particular small area.

Microclimates are of concern to the farmer, agronomist, forester, gardener, civil engineer, biologist, and many others. Microclimatology is the study of the effects of weathering in small areas. It is concerned with the heat exchange near the ground and with the various influences upon it.

A microclimate map made to study the agricultural use of land in a small area not far from Adelaide, South Australia, is given in Fig. 4 to illustrate some of the features of a microclimate.

2–3. The Five Variables of Weathering

In 1941 Hans Jenny gave the variables in weathering and soil formation as climate, parent rock, biological activity, topography, and time. The genetic or geographical relationships among soils are given as: $S = f(cl, o, r, p, t)$, where S is any soil property, f indicates a functional relationship, cl is climate, o is biological activity (organisms), r is topography with its effect on drainage, p is the parent material, and t is the time during which soil forms. Soil properties are related to climatic variables by the constants o, r, p, and t. This constancy is known as the steady state. The effect of the steady state in soil formation has been further explained by Jenny (1941) in terms of his five variables in terms of S–m functions shown by soils of localities with nearly the same mean annual temperature but considerable variation in mean annual rainfall. Soil property–moisture functions are $S = f(m)_{T,o,r,p,t}$ and soil property–temperature functions are $S = f(T)_{m,o,r,p,t}$, where m is the moisture variable and T is the temperature variable. Common manifestations are:

1. Decrease of calcium with increase of rainfall (more leaching, if drainage is good). However, the leaching of calcium may also be a time function as in the movement of calcium carbonate downward in loesses of different ages.
2. Increase of soil nitrogen and organic matter with increasing mean annual rainfall (increase of total vegetation).
3. Increase of clay content in soil with increase of mean annual temperature under which a rock weathers. Such an increase is partly caused by the fact that chemical weathering occurs in frost-free weather, and not under frozen conditions. Also, the kind of clay mineral formed differs because of increased weathering and leaching. Prolonged leaching results in the simplification of the kind of clay mineral formed. Thus, in a traverse from the north pole southward to the equator we find the following sequence:

   mica and mixed-layer mica–montmorillonite →
   kaolinite → gibbsite

   This clay mineralogy is an expression of the fact that leaching causes first the removal of cations and then silica, whereby the structures become simplified.

The fundamental differences in the mode of formation of soils of dry and humid regions are due to moisture and temperature.

Low Rainfall Areas. Water penetrates the soil to a limited depth; weathering and soil formation take place, but the products are not removed from the soil profile. The reaction (pH) is neutral or slightly alkaline. Water is removed from the weathering material by evaporation.

High Rainfall Areas. Water percolates through the soil and weathering material and is removed in groundwater. Soluble substances are removed and clay may be dispersed within the soil profile. Soils become leached and acid.

Temperature. High average temperatures in moist areas increase vegetation and rock weathering by chemical processes. High average temperatures in arid areas decrease vegetation and chemical weathering, and promote physical disintegration. Low average temperatures in moist areas cause soil freezing and permafrost, and generally slow soil-forming processes. Water movement is prevented or only occurs in a thin layer above the permafrost. Temperature is treated more extensively in Chapter 10.

2–4. The Steady State in Weathering

The steady state is the time–environment condition of a system that is open to its environment. Insofar as weathering is concerned, this means that a particular segment of the earth's surface would have to have been under the influence of the same kind of climate (and microclimate) for a very long time, generally to be measured geologically. A steady-state condition implies that the land surface has been influenced by yearly increments of precipitation and precipitated airborne chemical compounds that are deposited by rain. Variation in rainfall in different places has been shown by the reports of weather stations, and the work of the Meteorological Institute of Sweden is particularly important in showing seasonal variations in the chemical composition of rain over northern Europe.

It has long been known that in arid regions salts accumulate, but in humid regions atmospheric accessions are not so apparent. Figure 5 gives a diagrammatic picture of the main chemical processes and their products in various environments.

Investigations of the chemical composition of rainwater have stressed the importance of the geochemical cycling of certain chemical elements. A common illustration is the distribution of sea salts by rain. Eriksson

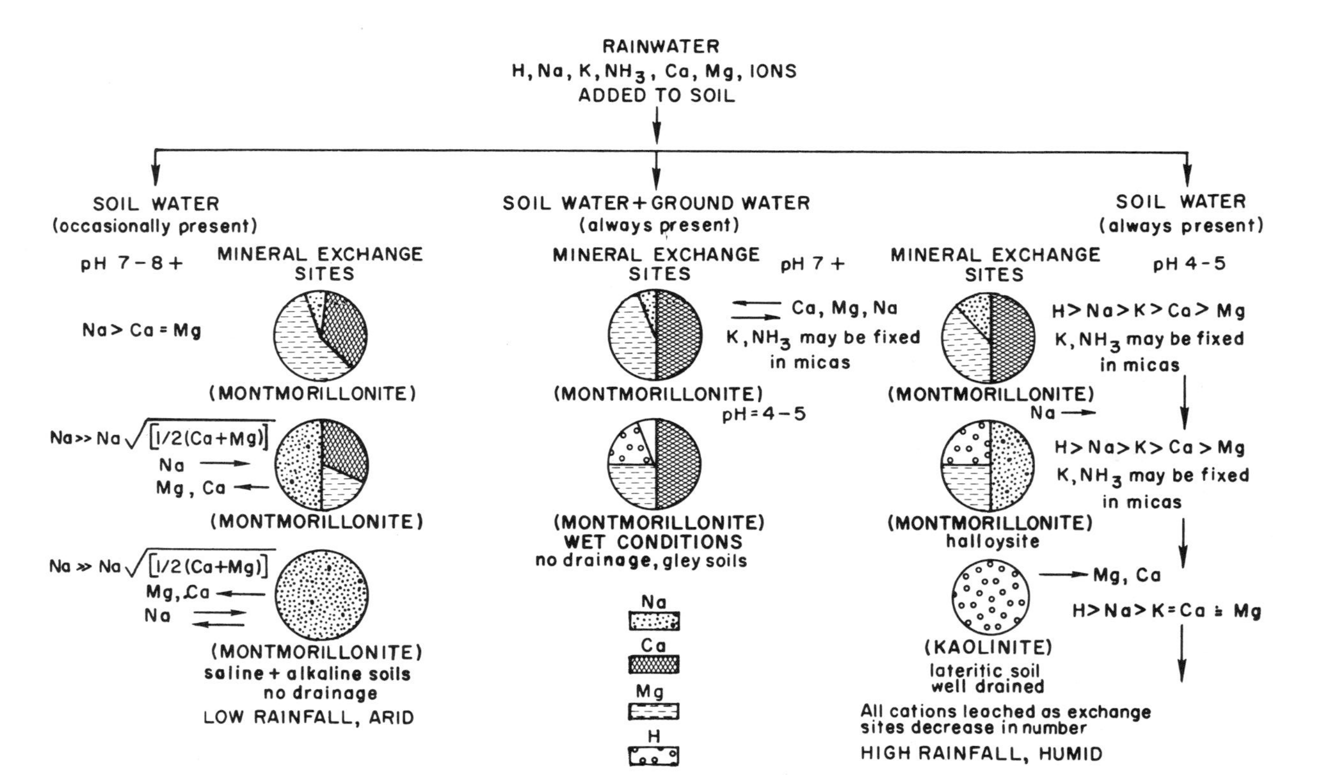

Fig. 5. Schematic representation of the relationship of rainwater to rock weathering and soil formation. Each circle represents exchange sites on clay mineral fully occupied by cations (after Carroll, 1962).

(1958) has summarized the long-continued chemical examination of the sodium chloride contents of rain in Western Australia whose climate is not influenced by mountain ranges that cause variations in the rainfall pattern.

The chemical climate is an important factor in weathering. After equilibrium has been attained, i.e., the steady state, the various acquired cations and anions in the system will be distributed in an orderly manner between the soil solution (the result of rainfall) and the mineral particles (the weathering rock), and the results of remaining in the same environment are expressed as soil variations on a large scale in climates and, on a small scale, as topographic and drainage variations due to microclimates. This aspect of weathering has a very important effect on ecology and on the biological circulation of elements. It is an essential part of geochemical circulation. Although both major and minor, or trace elements, are concerned in this cycling, it has been particularly studied for trace elements important in animal and plant nutrition. The chemical climate has been discussed in detail by Eriksson (1958).

Chapter 3

GEOCHEMICAL AND PEDOCHEMICAL WEATHERING

If a soild rock is broken down into a mass of rock fragments, the process is known as physical weathering (Chapter 7) and little change in the minerals of the rock takes place. In chemical weathering, however, the chemical elements in minerals (and therefore in rocks) are redistributed to a greater or lesser extent depending on the variables of weathering discussed in Section 2–3, and on the steady state discussed in Section 2–4. Chemical weathering of the outermost part of the lithosphere takes place in two stages; the first stage is the production of rotten rocks or saprolites, on which the second stage, soil formation, takes place. The first stage is geochemical weathering, and is mostly the inorganic alteration of solid rocks, but in the second stage the effects of vegetation, both living and dead, together with the effects of metabolism of microorganisms living in the geochemically altered rock material, are added to the continued inorganic processes, and this combination produces a soil which is the culmination of weathering at that particular locality.

3–1. Saprolites

In geochemical weathering the chemical alteration of minerals commences because of the chemical reactions that take place between minerals and their environment. A common example of weathering is seen in the change of feldspars into clay minerals. A weathered rock has a different appearance from that of a fresh rock. The constituent minerals (in an igneous or metamorphic rock) are dull, and they may be loosened from one another and/or from the matrix. The rock may crumble instead of remaining intact. The rock appears rotten and may be called a saprolite (Becker, 1895). The texture and structure of a saprolite are similar to those of the fresh rock and the saprolite can readily be recognized as a granite, basalt, gneiss, etc. There is little or no volume change or movement of alteration products. Leaching has changed feldspars to clay minerals and oxidation of ferrous iron to ferric iron has given the saprolite a brownish color. A saprolite is the product of chemical changes that have taken place *in situ* under continual moist conditions.

A saprolite that has developed from a Precambrian schist is shown in a road-cut in the Great Smoky Mountains in North Carolina in Plate 1B, and the fresh schist from which the saprolite formed is shown in a quarry face nearby in Plate 1C. The saprolite is similar to the schist in structure and texture except that oxidation of iron has produced a reddish-brown color. Closer inspection of the saprolite shows the hard minerals of the schist to have been largely replaced by clay minerals and that the saprolite is soft. It could be cut with a pick or spade when the road-cut was made, but exposure to the atmosphere has hardened it. Plate 1A shows a road-cut in a saprolitized gneiss from the same area. At this exposure geochemical weathering has gone a stage further than in Plate 1B and the deeply weathered gneiss has lost most of its original structure; the material has not yet developed a soil on top of the weathered zone.

In the tropics geochemical weathering is common and the saprolitic material may be up to 300 feet thick (Ruxton and Berry, 1957). Such a thickness of weathered rock *in situ* indicates lack of erosion that would remove it. Under tropical weathering conditions, as on the island of Kauai, Hawaii, where the rainfall is about 200 inches a year, the weathering rock is continually wet.

Saprolitization also occurs in rocks that are covered by later deposits that protect them from erosion. An example is saprolitization under a cover of river gravel or sand through which water percolates. Similar deep weathering occurs under snow cover that melts and slowly releases water to the underlying rock.

In Tasmania, Hale (1957) has recorded deeply weathered dolerite on the central plateau beneath glacial till. The rocks of the Australian Alps below a height of 5000 ft are saprolitic. Extensive areas in arid Western Australia are underlain by saprolitic rocks on which lateritic profiles have developed (probably in the Pliocene). These lateritic profiles were first described by Simpson (1912) and Campbell (1917). The original laterite described by Buchanan in 1807 in India is a saprolite that hardens on exposure to air. Terra rossa soils and bauxite associated with limestones are products of the saprolitizing process in easily soluble rocks. The red-earth residuals of Queensland (Bryan, 1939), the deeply weathered red soils of Guam (Carroll and Hathaway, 1963), and the latosols of Hawaii (Sherman, 1955) are all expressions of the same process.

Although the final expression of geochemical weathering is the production of a saprolite, there are many stages between slight and complete

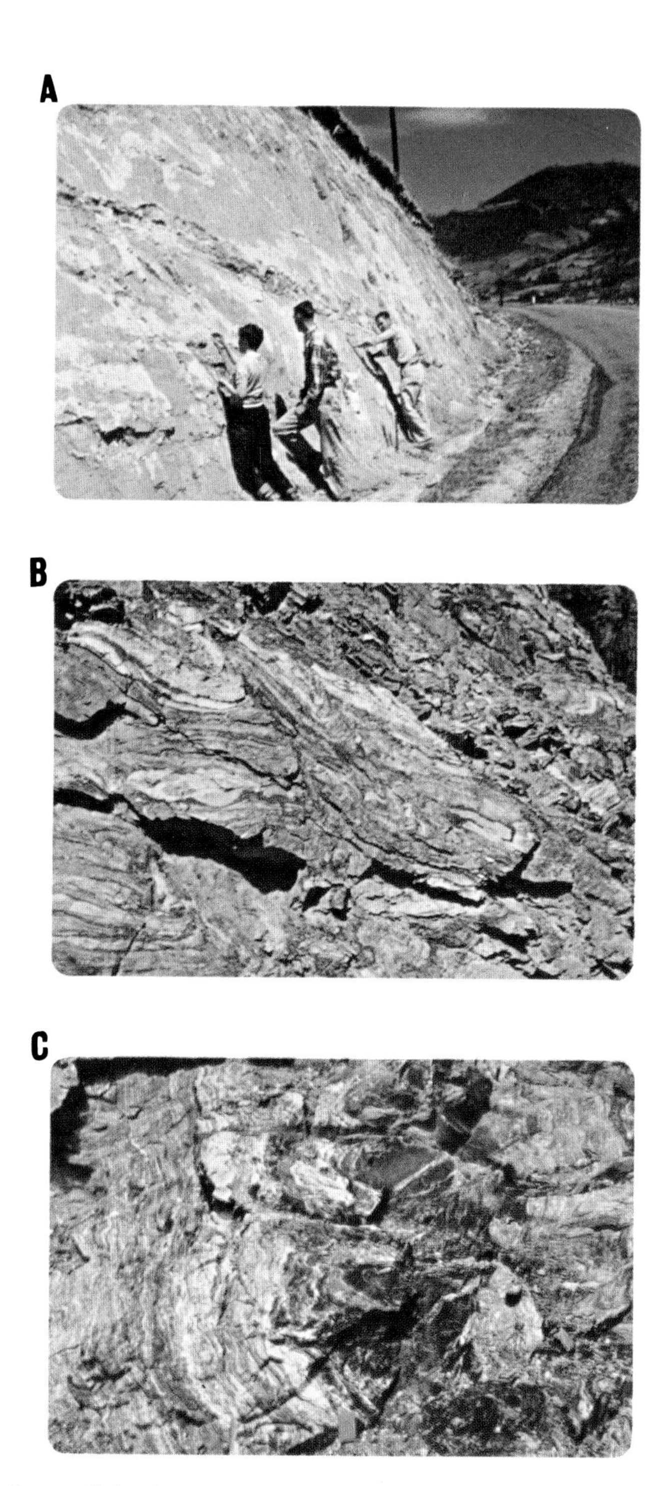

Plate 1. Deep soil developed from Precambrian schist, North Carolina. (A) Strongly weathered (saprolitized) gneissic schist with soil surface, North Carolina. (B) Saprolite developed from Precambrian schist, North Carolina. (C) Schist from which saprolite B has developed (photos, D. Carroll).

alteration that can be recognized. The effectiveness of geochemical weathering varies with the climate and, locally, with the interaction of Jenny's four other variables—parent material, biological activity, topography, and time. A general, overall outline of the effect of these variables is given in Table 3.

Probably the most extensive area of saprolitization now preserved on the earth's crust is that in Western Australia. Prider (1966) has summarized the information describing the area covered by a laterite crust and the kaolinized rock beneath it. Saprolitization (expressed by kaolinization) occurs to a depth of 100 ft or more below the laterite capping over an area of about 300,000 square miles. The saprolite is preserved because it is present on a very slightly dissected plateau in a dry region where erosion is at a minimum. In this saprolite the original structure and texture of the Precambrian gneisses, granites, schists, and dolerites remain, and it is possible to make geological maps as though the terrain were underlain by fresh rocks. The formation of saprolite was on a regional basis in a climate that produced forest vegetation and had sufficient rainfall for continuous leaching. Saprolitization occurred between late Miocene and middle Pliocene times.

3–2. Soils

The term *pedochemical weathering* is used to describe the alterations that take place in geochemically weathered material that lead to the formation of a soil. It has been well said by Berzelius that "The soil is the chemical laboratory of Nature in whose bosom various chemical decompositions and synthesis reactions take place in a hidden manner." Soils have been investigated scientifically for over 100 years, and the greatest advances in our knowledge of soil formation have come from government research institutions whose responsibility has been to assist agricultural practices. Such research has been necessary to increase the production of food for mankind. During the 20th century agricultural research has found chemistry and modern chemical technology and instrumentation of paramount importance in the study of soil-forming processes.

Pedochemical weathering results from biological activity whereby organic matter and humus are added to, and react with, the mineral constituents of geochemical weathering. Thorough vegetation pedochemical weathering is primarily dependent on climate. Different types of vegetation can grow in the various climatic zones. The vegetation produces humus in various quantities and the products of humus enter into organic chemical reactions with the cations (principally) and the anions (slightly) of the weathered minerals, and the mineral alteration phases, of the parent rock. The result of these reactions is a soil profile (Section 4–2) which, as a result

TABLE 3
Factors of Variability of Weathering

Climate	Parent material*	Biological activity	Topography†	Time
Polar Arctic Antarctic Temperate dry humid Subtropical dry humid Tropical dry humid	(*a*) Igneous and metamorphic rocks: granite, gabbro, basalt, diabase, gneiss, schist (*b*) Sedimentary rocks: sandstone, shale, limestone, dolomite	Polar nil ↕ Tropical maximum	Flat and sea level; flat at various elevations; undulating; rugged	
Produces circulating water	Produces chemical and mineralogical differences		Causes changes in ground water and rivers; in drainage generally	Causes maximum or minimum completion of the end point for the interaction of variables

* Variations due to disequilibrium (a) and equilibrium (b).
† Causes catenas of soil types on similar rocks and microclimatic variations.

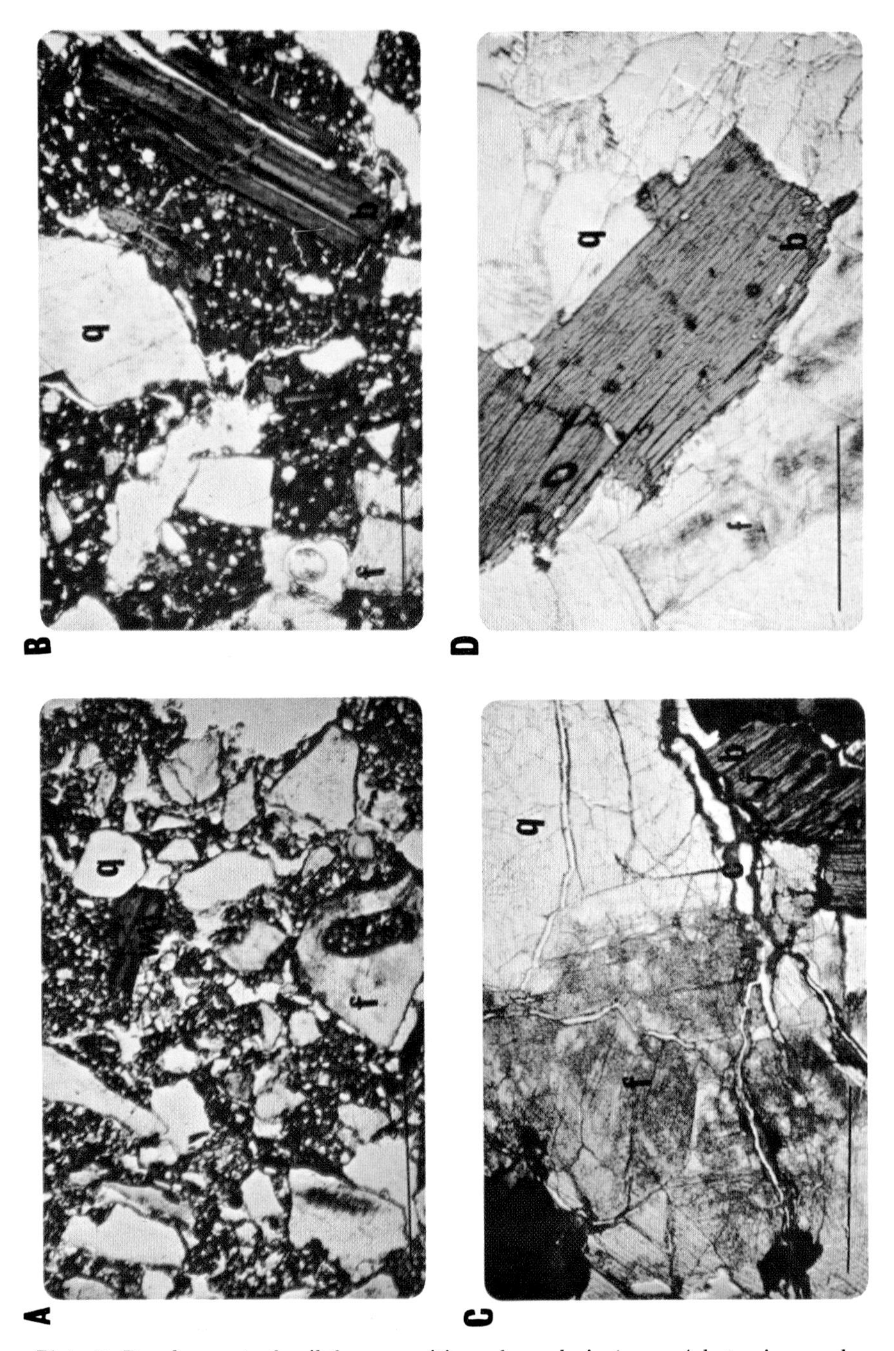

Plate 2. Development of soil from granitic rock; scale is 1 mm (photomicrographs, unpolarized light, Roy Brewer, C.S.I.R.O. Canberra, Australia). (A) A_2 horizon, 14 cm below soil surface, showing quartz (q) and clay-sized material with organic matter disseminated throughout; (f) feldspar; biotite altered to clay (wb). (B) B_2 horizon, 30 cm below surface, showing quartz (q), feldspar (f), a large flake of biotite (b) in an iron-stained matrix. (C) Weathered rock, 43 cm below surface, showing quartz (q), feldspar (f) slightly altered, biotite (b) in large oxidized grains, and a crack with illuviated clay (c). (D) Granitic rock with quartz (q), feldspar (f) with incipient alteration in places, and biotite (b). This rock is the parent material of soil A.

of steady-state conditions, is characteristic for the climate zone, and forms the basis of soil mapping.

Pedochemical weathering varies in its intensity according to the weathering environment (Chapter 2). When pedochemical weathering is at a minimum the mineral content of the upper part of the soil is very similar to that of the comminuted rock from which it developed. Such soils are poorly developed and are known as skeletal soils or lithosols. Often the weathered material has not actually formed a soil. The soil map of the United States (Fig. 10, Section 5–3) shows large areas of lithosols in the Rocky Mountains area.

In situations in which pedochemical weathering is clearly evident (Jackson and Sherman, 1953), the following has occurred:

1. Release of anions and cations from minerals in the soil, including those taken up by plants and those lost by leaching.
2. Deposition of sesquioxides and clay formation in the B horizons (the B horizon is generally considered to be the position in the profile where the strongest pedochemical reactions take place).
3. Pedochemical accumulation of sesquioxide and titaniferous surface horizons and crusts (certain soils in Hawaii exhibit titaniferous surface horizons).
4. Coincidence of the depth of chemical weathering with the depth of biological activity and soil development.

Ionic release is the first step in weathering reactions that lead to the formation of new mineral products. Chemical weathering as a part of the process of pedochemical weathering causes the development of the colloidal part of soils that is largely made up of clay minerals.

The change from a fresh granite into a soil is shown in a series of photomicrographs in Plate 2. The parent rock is a granite consisting of quartz, feldspar, and biotite mica (Plate 2D). As the rock weathers geochemically it appears as in Plate 2C, in which the feldspar has developed sericitic patches and the biotite has turned brown from the oxidation of ferrous iron in the biotite crystals to ferric iron. Some of the ferric iron has been mobilized and has moved into cracks in both the quartz and the feldspar. In Plate 2B the mineral grains of the rock are no longer interlocking as they were in the fresh granite. The feldspar has largely disappeared, its place being taken by a ferruginous matrix consisting largely of clay. The biotite has a less robust character, and is more iron-stained in Plate 2C. The material is a soil, the result of pedochemical weathering induced by the growth of plants on the geochemically weathered rock. Plate 2A shows the soil near the surface. The quartz grains have been broken into various sizes, the initial breaks starting in the cracks in the fresh quartz that are clearly

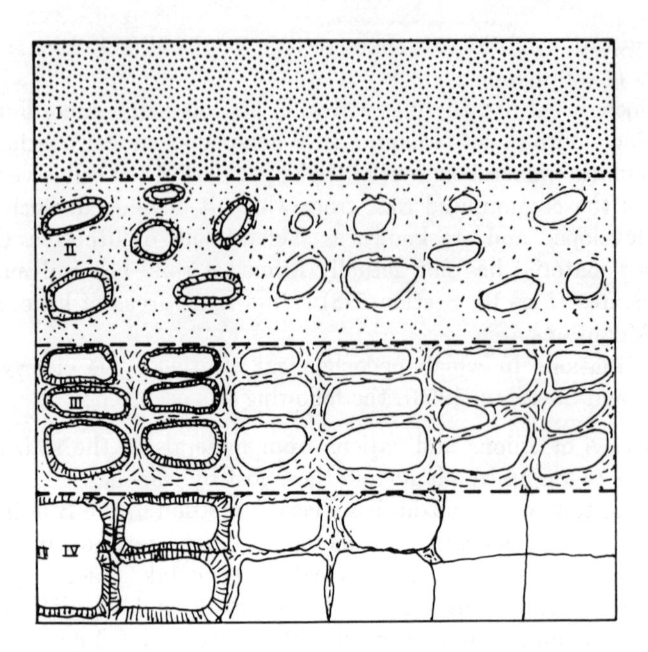

Fig. 6. Weathering of granite in Hong Kong. I) Residual debris, structureless sand, clay, or clayey sand. II) Residual debris with core stones; core stones subordinate, rounded, and free. III) Core stones with residual debris; core stones dominant, rectangular, and locked. IV) Partially weathered rock, minor residual debris along major structural planes, but may be considerably iron-stained (iron staining shown inside core stones at left of diagram only) (after Ruxton and Berry, 1957).

shown in Plate 2D. The clay matrix has increased and is now the dominant part of the soil. Feldspar has been almost completely removed and only the largest grains remain. The biotite is only barely recognizable as a dark-brown flake with prominent cleavage lines. It has changed its mineralogical and optical character, and now can only be called weathered biotite. This weathering sequence is only about 50 cm deep.

In contrast deep weathering, both geochemical and pedochemical, has been described in Hong Kong by Ruxton and Berry (1957). A number of different stages in this weathering could be recognized, some of which were due to erosion of the weathered rock and some caused by soil creep. The major development of the weathering and its variations is shown diagrammatically in Fig. 6.

Chapter 4

SOIL

The uppermost part of the weathered rock at the earth's surface is called soil. In "Soil Classification, A Comprehensive System, 7th Approximation" published by the U.S. Department of Agriculture in 1960 there appears the following definition of soil. It is similar to that given in Chapter 3 when describing pedochemical weathering:

> "Soil . . . . is the collection of natural bodies on the earth's surface, containing living matter, and supporting or capable of supporting plants. At its upper limits is air or water. At its lateral margins it grades to deep water or to barren areas of rock, ice, salt, or shifting desert sand dunes. Its lower limit is perhaps the most difficult to define. Soil includes all horizons differing from the underlying rock material as a result of interactions between climate, living organisms, parent materials and relief. Thus, in the few places where it contains horizons impermeable to roots, soil is deeper than plant rooting. More commonly soil grades at its lower margin to hard rock or to earthy materials essentially devoid of roots. The lower limit of soil therefore is normally the lower limit of the common rooting of the native perennial plants, a diffuse boundary that is shallow in deserts and tundra and deep in the humid Tropics."

4–1. Soil as an Entity

A soil is not a discrete individual like a plant or animal, and is not clearly separated from another soil into which it generally grades gradually on its margins. The smallest volume of material that can be called a soil is the pedon (Greek *pedon*—ground). A pedon is a three-dimensional unit of material that is characteristic of the weathering at a particular locality; it is a unit that can be distinguished in the field by physical characteristics such as color, texture, and arrangement of its parts, so that it forms a mappable piece of ground that can be separated from other mappable units.

If a vertical cut is made from the surface of the soil to the underlying rock, a horizontal arrangement of layers can generally be seen. This section through the soil is known as the *soil profile;* the individual layers are the *soil horizons.* Each horizon may differ from its neighbor in color and size, and in physical and chemical composition. The arrangement of the layers forms the basis of soil classification; soils should be properly studied in the field, because if brought into the laboratory their horizons are destroyed and they lose their identities, although many of their characteristic features such as grain size and amount of clay present can be examined. Through careful observations by many soil scientists a number of parts, or horizons, can be classified in a soil profile. A hypothetical soil profile with all the different kinds of horizons is shown diagrammatically in Fig. 7. Natural soil profiles do not necessarily have all these horizons. The arrangement differs according to the climate in which the soil developed and according to the parent rock below the soil.

Originally the A, B, C names for soil horizons were applied by Russian soil scientists to horizons they found in black soils called chernozems. The A was used for the uniformly dark-colored surface soil, the C for the weathered material below the soil proper (solum), and the B for the transitional horizon between them. Russians first studied soils in the field in order to classify them for land use. Later these designations were used on ashy gray to white soils called podzols and on other similar soils in northern Europe. The B horizon became the layer of accumulation as shown in Fig. 7. The appearance of a podzol is shown in Plate 3A. The soil profile clearly shows a bleached layer, A, overlying a yellowish layer, B.

4–2. Soil Profiles

A soil that represents the maximum pedochemical weathering in any region (a mature soil) is the chemical and physical expression of the chemical environment produced by the climate acting on a certain type of rock. Because there is a sameness in the pedochemical results in a region, the arrangement of the horizons in a soil profile are similar. The same kind of material weathering in a different climate will likewise form a profile of its chemical environment. In Fig. 8 eight hypothetical soil profiles are given to illustrate the variation in horizons in the profiles and the nomenclature that has been given to the horizons by the U.S. Department of Agriculture.

Figure 8 gives diagrams of the kind of soil profiles developed in the main groups of soils common to the various weathering situations. In Fig. 8A the soil is a gray-brown podzolic; B, a chernozem (black earth); C, a latosol (lateritic soil); D, a solodized solonetz (a salty soil with

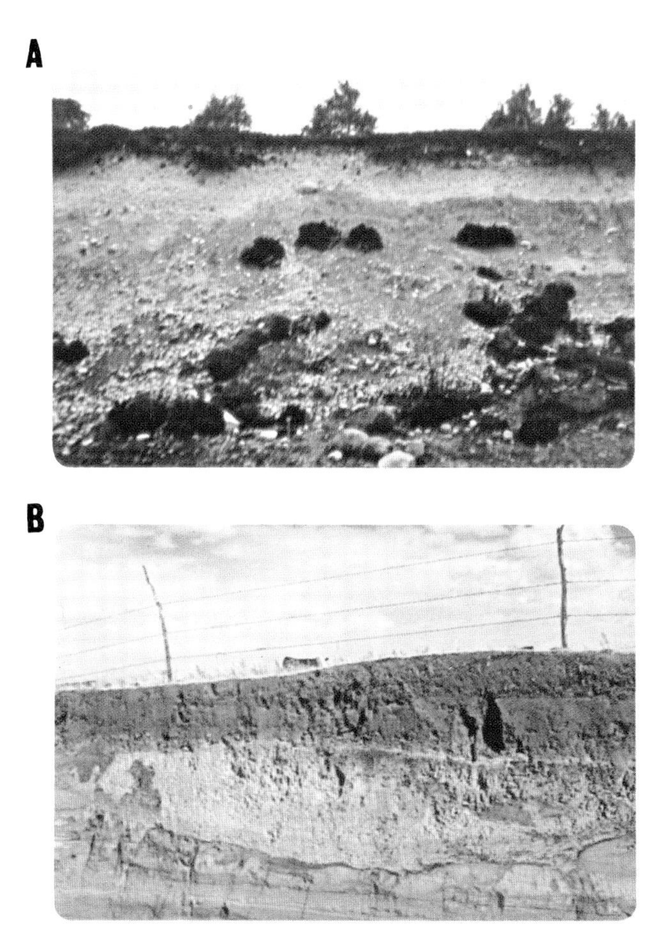

Plate 3. Types of soils. (A) Profile of a podzol in northeastern Scotland. (B) Fossil soils exposed in a road-cut in northwestern Kansas (photos, D. Carroll).

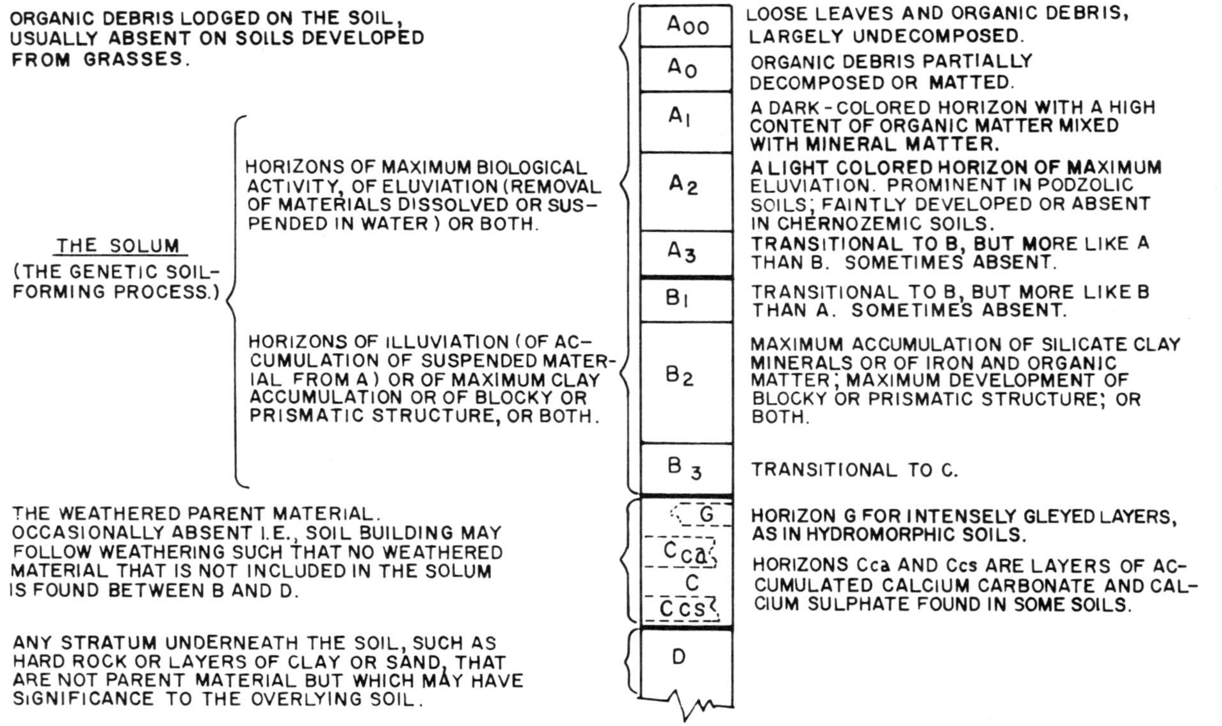

Fig. 7. Diagram of a hypothetical soil profile showing the arrangement of horizons (Soil Survey Manual, U.S. Dept. Agriculture, 1951).

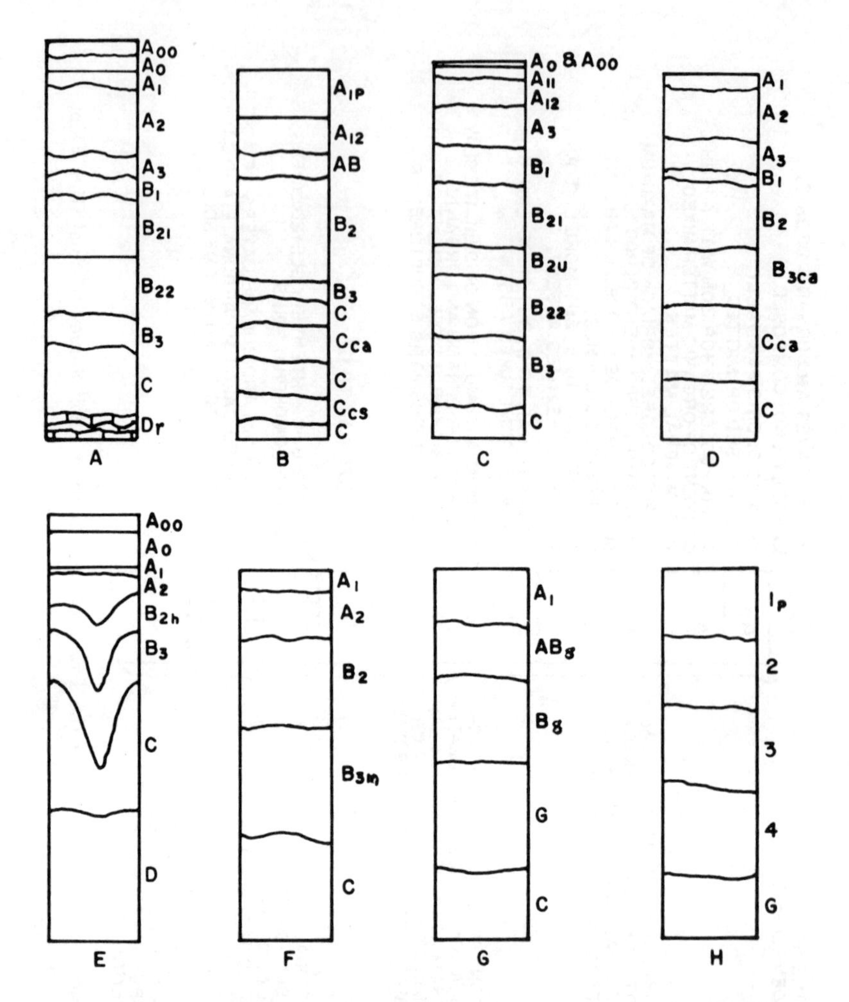

Fig. 8. Arrangement and nomenclature of soil horizons developed under different climates (Soil Survey Manual, U.S. Dept. Agriculture, 1951).

columnar structure); E, a podzol; F, a planosol (a strongly eluviated interzonal soil); G, a humic gley (waterlogged soil); and H, a bog soil. All these soils are strongly leached with the exception of B (a pedocal) and D, G, and H which are caused by local drainage conditions (see Section 5–2). In Fig. 8 some of the horizons characteristic of the soil profiles have received letter designations in addition to the A, B, C, and D of Fig. 7.

These letters have the following meanings in profile descriptions:

- ca, calcium carbonate accumulation;
- cs, calcium sulfate (gypsum) accumulation;
- u, unconformable layer with inherited characteristics different from those of the adjacent soil horizon, e.g., a stone line;
- h, accumulation of decomposed organic matter;
- m, indurated layer composed mainly of silicate minerals;
- g, a greyed horizon (a layer of low redox potential in which Fe^{+3} is reduced to $Fe,^{+2}$ see Section 8–6);
- p, a disturbed layer, as in ploughing.

The A, B, C, D nomenclature of soil horizons has been applied mainly to profiles consisting of mineral matter. The separate layers found in some profiles are subdivided by the use of subscript numbers, as B_{21}, B_{22}, and B_3 in Fig. 8A. In profiles consisting predominately of organic matter the horizons are designated 1, 2, 3, 4, as in Fig. 8H.

As a soil profile develops in weathered material colors are formed in the different horizons as a result of chemical weathering. These colors are characteristic of the kind of chemical weathering. The most easily recognized color change is that of the oxidation of iron which produces yellow and red colors, and of the reduction of iron which produces bluish gray colors. Colors do not appear similar to all persons; in order to facilitate descriptions of soil profiles that are meaningful, all soil colors in the United States are compared to those of the Munsell Soil Color Chart which was largely evolved by scientists in the U. S. Department of Agriculture. Soil colors are of great importance in soil profile descriptions, and an example of their use is given in the description of the Clarksville series of red-yellow podzolic soils that develop from the weathering of the cherty limestone ridges of northern Alabama, northwestern Georgia, eastern Tennessee, southwestern Virginia, and the Ozarks region of Arkansas and Missouri.

Soil Profile: Clarksville cherty silt loam—forested

A_1	0–1″	Dark grayish brown (2.5Y 4/2) or grayish brown (2.5Y 5/2) cherty silt loam with weak fine granular structure; very friable; strongly acid; boundary abrupt, smooth. ·½ to 3 inches thick.
A_2	1–6″	Pale brown (10YR 6/3) cherty silt loam with weak fine granular structure; very friable; very strongly acid; boundary gradual, smooth. 3 to 7 inches thick.
A_3	6–14″	Light yellowish brown (10YR 6/4) cherty silt loam with weak fine subangular blocky macrostructure and weak fine granular microstructure; very friable; very strongly acid; boundary gradual, wavy. 6 to 8 inches thick.

B_1	14–21″	Yellowish brown (10YR 5/4) to brownish yellow (10YR 6/6) cherty or very cherty silt loam with weak fine and medium subangular blocky structure; friable; very strongly acid; boundary gradual, smooth. 5 to 10 inches thick.
B_2	21–38″	Yellowish brown (10YT 5/6) or strong brown (7.5YR 5/6) very cherty silt loam or light silty clay loam with few fine faint variegations of yellowish red and pale brown in the lower part; weak to moderate medium subangular blocky structure; few faint clay films; friable; very strongly acid; boundary gradual, wavy. 10 to 24 inches thick.
C	38–72″+	Chert bed, with interstices filled with yellowish brown silt loam or silty clay loam variegated with yellow, red, and gray; the chert is white and yellowish brown; some is weathered to yellowish red or red. Several feet thick.

Topography: Chiefly steep side slopes of the higher ridges, with small areas of sloping ridge tops that are generally broader and proportionately more extensive in the Ozark plateau than elsewhere in areas of Clarksville soils. Gradients range from about 4 to 40%; 10 to 25% probably is the dominant range.

Drainage and Permeability: Well drained to excessively drained. Runoff is medium to rapid and permeability is moderate to rapid.

Vegetation: Originally hardwood forests; chiefly red, black, white, Spanish, scarlet, blackjack, and post oaks; hickory, tulip poplar, elm, chestnut, ash, sweet and black gum, persimmon, and dogwood. Second growth includes many Virginia, shortleaf, and loblolly pines and low-grade hardwoods.

Use: Only partly cleared; many areas previously cleared are idle or in second-growth timber. Cleared areas are used for a wide variety of crops and pasture. Crops include cotton, corn, small grains, cowpeas, soybeans, sorghum, hay, and, in places, vegetables, melons, small fruits (including strawberries), peaches, apples, and Burley tobacco.

Type Location: Bledsoe County, Tennessee, 1.5 miles east of College Station.

Series Established: Montgomery County, Tennessee, 1901.

4–3. Mineral Products of Weathering

A soil consists of mineral particles of various sizes, organic matter, living organisms, water, and air. These materials give soils their physical and chemical composition. How these materials are arranged within a soil profile is a function of both macroclimate and microclimate, and depends on the length of time the soil has been situated in its present position. The mineral particles are derived from the disintegration of the parent rock; but material may be added to the developing soil as volcanic dust or loess. Some minerals do not alter easily, and so they remain as discrete grains in the soil. Other minerals are altered to clay minerals, and these are mainly responsible for soil fertility. Clay minerals are extremely fine grained, and the clay fraction is often referred to as "soil colloid." The standard size of clay particles is generally taken as less than 2 μ in diameter. Further details concerning the grain size of soils are given in Section 7–4.

Any rock, provided it contains weatherable minerals, can provide the material for soil formation. Rocks can be classified into three principal kinds: igneous, sedimentary, and metamorphic.

Igneous rocks are formed by crystallization of magma either well below the earth's surface (granite), or at the surface (basaltic lava flows). The chemical composition of these rocks depends on that of the parent magma. Their grain size depends on how the liquid material (magma) crystallizes. An outline of the composition of igneous rocks is given in Section 1–1.

Sedimentary rocks consist of materials removed from other rocks by erosion or those chemically accumulated with or without the presence of organisms. The materials of sedimentary rocks are transported by water to a marine environment where they accumulate as marine sediments. Such sediments are derived from a land surface that is being actively eroded. A step in the erosion process very often is geochemical weathering, followed by pedochemical weathering in temperate and humid climates; all the processes of chemical weathering and accumulation of mineral grains and the winnowing of chemical constituents take place resulting in a modified material that is the parent material of the sediment. In cold climates erosion may be by glacial scour resulting in rock flour and in chemically unaltered rocks of moraines. Nivation results in splintery fragments of unaltered minerals or chips of rock. In tundra climates the effect of periodic leaching with organic acids of peaty waters removes soluble minerals, leaving an accumulation of the more resistant grains originally present. Such leaching occurs in the warmest part of the year above the permafrost. In the marine environment, depending on the temperature of the water, the comminuted detritus from the land surface may become mixed with calcium carbonate and a limestone may result. Shales may or may not contain calcium carbonate, but the detritus is more important than the calcareous ingredient. The limestones of Guam are an example of the mixture of a small amount of volcanic material in an environment of calcium carbonate deposition (Hathaway and Carroll, 1964). With greater quantities of volcanic material added to the sedimentary environment, tuffs result.

Sedimentary rocks fall into three principal groups: mechanical, chemical, and mixtures of mechanical and chemical. The mechanical group may be either coarse or fine grained, depending on the agents of erosion and the transportation of the eroded material. The coarse-grained rocks are conglomerate, breccia, gravel, medium-grained sand, arkose, grit, graywacke, and sandstone. The fine-grained rocks are mudstone, shale, siltstone, loess, unconsolidated silty clays, and clays. The principal chemical sedimentary rocks are the limestones and chalks. In some geological situations siliceous spongolite and diatomite occur and are important. Sedimentary rocks that are mixtures of mechanically disintegrated rock material and chemical deposition are calcareous shales grading into shaly and impure limestones, calcareous tuffs, limy clays, calcareous graywackes, and many others.

Metamorphic rocks consist of original minerals of igneous and sedimentary rocks that have been altered by heat or pressure or both. Typical examples are gneisses and schists. For example, pressure and heat applied to mud results in the formation of this sequence of rocks:

mud → shale → slate → phyllite → mica schist → garnet–mica schist → kyanite–mica schist → sillimanite–mica schist

The effect of metamorphism on a rock depends primarily on the ability of the minerals present to be changed by pressure or heat. For example, a rock containing only quartz cannot be materially altered by metamorphism.

The principal petrographic features of the main types of rocks available to act as parent materials for soils have been outlined to give background information (for further details petrographic textbooks should be consulted; one such is *Petrography*, by Williams, Turner, and Gilbert, Freeman, San Francisco, 1958).

The important minerals of a rock from a soil-forming point of view are:

Quartz:	only slightly weatherable; forms the skeleton of the soil
Feldspars:	weatherable, form the clay minerals of the soil
Pyroxenes: *Amphiboles:* *Micas:*	weatherable to different degrees, form the clay minerals and iron oxides of soils
Accessory minerals:	very stable; persist as individual grains in the sand and silt fractions of soils. As many contain traces of rare elements these minerals, on slow weathering, act as a reserve or "bank" of certain elements that can be "drawn upon" during

geologic time, or as the soil is used for agricultural, pastoral, or forestry purposes. One such mineral is tourmaline which releases boron, an essential chemical element for plant growth (see Section 12–7)

The association of chemical elements with minerals can be arranged in a scheme that is similar to one that was originally published in Dutch by Mohr and translated into English in 1944. This scheme is given in Fig. 1 (Section 1–1).

4–4. Residual Minerals of Weathering

Minerals have different inherent tendencies to alter by weathering processes at different rates that are partly due to the structure and composition of the minerals and partly to the weathering environment. A weathering stability series was formulated by Goldich (1938) and forms the basis of much of our thought concerning the stability of minerals in the weathering environment (Fig. 9).

The amount of residual minerals in soils is often considerable. The amount varies according to the above stability series, but the contribution of these minerals to weathering is due to the kind of parent rock that supplies them. Other factors causing variation in the presence of residual

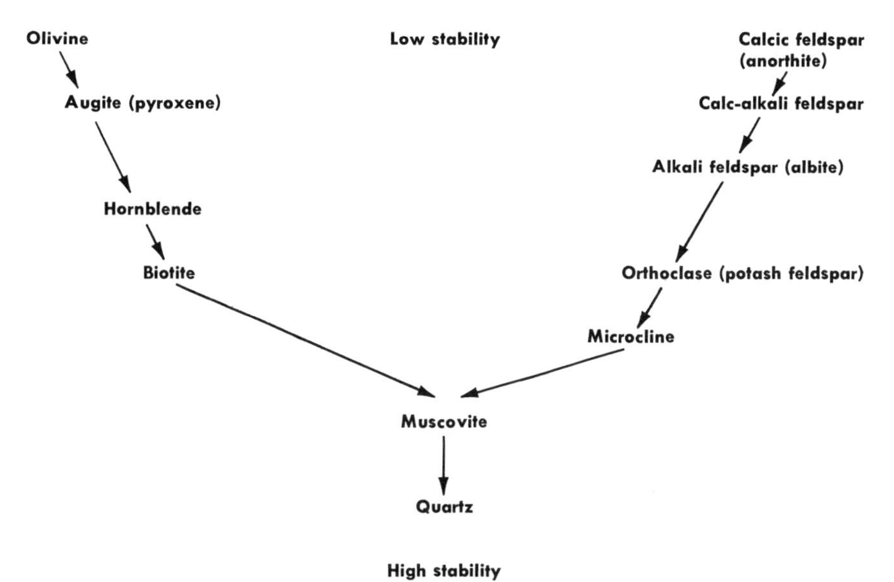

Fig. 9. A weathering stability series (after Goldich, 1938).

TABLE 4

Amounts of Residual Minerals in the Fine Sand of the Ploughed Layer of Some Pennsylvanian Soils (after Jeffries, 1937)

Minerals	Lackawanna sandy loam	Hagerstown silt loam	Volusia clay loam	Dekalb A sand	Dekalb B clay loam
Quartz	361,549	243,842	383,130	189,000	479,900
Feldspars	2,264	38,666	36,000	21,536	5,034
Hornblende	761	140	93	—	52
Chlorite	2,174	214	1,405	—	47
Muscovite	2,215	391	1,346	386	114
Tourmaline	788	329	1,033	173	228
Iron oxides	6,900	1,816	2,685	1,025	2,523
Rutile	—	—	—	—	90
Zircon	761	679	1,837	478	1,683

All figures in lb/acre.

minerals are the climatic and topographic conditions under which a rock forms soil. The grain-size analyses of soils indicate in a general way the amount of residual rock fragments (large) and mineral grains (small) in weathering products (Section 7–4). More detailed information about the quantities of minerals in the fine sand (0.2–0.05 mm grain diameter) in some Pennsylvanian soils was obtained by Jeffries (1937). These soils were derived from limestones from which the calcium carbonate was removed during weathering. The residual minerals represent detrital minerals that were sedimented in a limestone environment. Some of the feldspars, however, may have formed authigenically in the limestones. Table 4 gives the residual minerals found in certain limestones by Jeffries. The concentration of minerals resistant to alteration during soil formation is a consequence of the range of stability found in the common rock-forming minerals.

During weathering it has been found that certain elements are removed from rocks and minerals in this order: Na, Ca, K, Mg, Si, Al, and Fe. From this Reiche (1950) proposed a weathering potential index, WPI, for rocks. It is the mole percentage ratio of the sum of the alkalis and alkaline earths, less combined water, to the total moles present, exclusive of water,

$$\frac{100 \times \text{moles}\ (K_2O + Na_2O + CaO + MgO - H_2O)}{\text{moles}\ (SiO_2 + Al_2O_3 + K_2O + Na_2O + CaO + MgO - H_2O_2)},$$

and is calculated from a chemical analysis of a mineral or rock.

Minerals of low stability have a high WPI. This formula can be used

TABLE 5

Weathering Potential Index for Some Common Rocks and Minerals

Material	WPI average	WPI range
Olivine	54	44–65
Augite	39	21–46
Hornblende	36	21–63
Biotite	22	7–32
Muscovite	10	
Labradorite	20	18–20
Andesine	14	
Oligoclase	15	
Albite	13	
Quartz	1	0–1
Granite	7	
Orthoclase	12	
Basalt	20	

for all rocks and minerals for which chemical analyses are available. Table 5 gives the WPI for some common rocks and minerals.

According to Reiche's formula both microcline and orthoclase (potash feldspars) have the same WPI, but it is well known that microcline is the more stable of the two.

The presence of certain minerals of the WPI scale in a soil indicates the kind of rock that is the parent of the soil (Section 1–1) and the completeness or maturity of the weathering processes in the development of a complete soil profile, that is, the zonal soil profile that is characteristic of the climate in a particular locality. Thus, if in a soil developed from basalt grains of olivine are observed in the fine sand fraction, the soil is immature and the profile incomplete, as olivine has a WPI of 54 and is the least stable of the common rock-forming minerals. However, if no olivine grains are present and only a few augite grains (WPI 39) can be recognized, it is concluded either that weathering has continued for a long time, probably at least 30% longer than if olivine alone had been removed, or that leaching was more thorough and more continuous. As in all weathering and soil-forming processes a number of variables operate together to produce the soil profile that is seen in the field.

4–5. Clay Minerals

The finest, often colloidal, part of the weathering material consists largely of clay minerals. These minerals result from the leaching of rock-

TABLE 6

Classification of the Clay Minerals (after Brindley et al., 1968)

Structure	Minerals	Remarks
1:1	Kaolinite	
	Dickite	
	Nacrite	
	"Fire-clay mineral"	Disordered kaolinite
	Metahalloysite ($2H_2O$)	Dehydrated; $d(001) = 7$ Å
	Halloysite ($4H_2O$)	Contains interlayer water; $d(001) = 10$ Å
	Allophane	Unordered or slightly ordered grouping of SiO_2, Al_2O_3 with H_2O
	Mica group	
2:1	Muscovite, 2M	Dioctahedral; layers bonded together with K^+ ions; generally detrital
	Illite	Dioctahedral; layers partially bonded together with K^+ ions; common mica of soils and sediments
	Glauconite	Marine sediments; somewhat like illite; mostly diagenetic
	Celadonite	Vugs in basalt
	Biotite	Trioctathedral; alters readily
	Vermiculite	Trioctahedral
	Montmorillinite group	
	Many species	Units separated by a water layer containing exchangeable cations that neutralize the structure; montmorillonites expand in organic liquids
	Chlorite group	Trioctahedral
	llb polytype	Common detrital variety
	lb, polytype	Diagenetic variety, low energy form
	Chlorite	Dioctahedral
	Chlorite, swelling	Corrensite, sudoite
	Chamosite	Generally a 7 Å chlorite
Chain silicates	Sepiolite	
	Palygorskite (attapulgite)	
Mixed-layer clay minerals	Rectorite	Regular; give an integral series of basal spacings from a long $d(001)$ spacing
	Montmorillonite-chlorite	
	Corrensite (see chlorite)	
	Mica-chlorite	Random; do not give an integral series from $d(001)$
	Vermiculite-chlorite	
	Mica-vermiculite	

TABLE 7
Principal Clay Minerals Found in Major Soil Groups (after Toth, 1964)

Great soil group	Principal clay mineral
Tundra	Illite
Desert	Mixed-layer, montmorillonite
Red desert	Illite
Chestnut	Montmorillonite
Chernozem	Montmorillonite, mixed-layer
Regur, India	
Margalitic soils, Indonesia	
Black turf soils, Africa	Montmorillonite
Black earths, Australia	
Dark magnesium soils, Hawaii	
Prairie	Montmorillonite, mixed-layer
Noncalcic brown or shantung	Illite, mixed-layer
Podzol	Illite
Gray-brown podzolic	Illite, kaolinite
Red-yellow podzolic	Kaolinite
Laterites	Kaolinite
Tropical, nonlateritic	Halloysite, montmorillonite
Solonchak	Similar to surrounding soil type
Solenetz	Similar to surrounding soil type
Solodi	Variable
Humic, gley, wiesenboden	Montmorillonite
Planosols	Illite, mixed-layer, kaolinite
Rendzina	Montmorillonite
Alluvial	Similar to surrounding soil type
Loess	Illite, mixed-layer

forming minerals by dilute aqueous solutions of varying composition and pH. The clay minerals are hydrous aluminum silicates that are extremely small crystals. The structural units of the clay-mineral crystals are silica tetrahedra and alumina octahedra that are bonded together into bands or layers. There are three ways in which these layers are combined to produce minerals:

1. As one layer of silica tetrahedra and one layer of alumina octahedra. These minerals are known as 1:1 layer silicates; kaolinite is the commonest example.
2. As one layer of alumina octahedra sandwiched between two layers of silica tetrahedra. This structure is a 2:1 layer silicate, a common example of which is mica.
3. The silica tetrahedra and alumina octahedra form a chain structure as in attapulgite (palygorskite). In the chain structure there are open

channels which allow water and other cations to enter. As the channels have a certain diameter only cations of that size, or smaller, can enter. Cation occurrence in chain silicates is similar to zeolite occurrence.

The common clay minerals belong to either 1 or 2 above. The chain silicates occur in more specialized situations, but do occur in some desert soils. A classification of the principal clay minerals is given in Table 6.

Identification of the clay minerals in soils of the earth's outer crust has been made possible by x-ray diffraction techniques. Prior to World War II, although the principals of x-ray diffraction were well known and had been investigated many years earlier, x-ray diffraction units had not been perfected or produced commercially and were not available to research institutions or to universities. From about 1930 a few clay-mineral investigators were able to use x rays in the identification, but the majority had to depend on the physical character and behavior of clay minerals under different natural and laboratory situations. Clay mineralogy is a mid-20th century study that was made possible by the development of suitable x-ray instrumentation. As understanding of the structure of clay minerals has progressed, new techniques for their identification have been found. Most of the investigations have been made by dedicated scientists in mineralogy and soil science, and among them the names of C. S. Ross, S. B. Hendricks, G. W. Brindley, and R. E. Grim are outstanding, although numerous other scientists should also be mentioned.

Because of the nature of clay minerals and their development from the rockforming minerals (largely feldspars) under varying natural weathering conditions it is possible to generalize concerning the kind of clay mineral associated with certain major groups of soils. One such grouping of clay minerals with soil types is that of Toth (1964) given in Table 7.

Chapter 5

SOIL PATTERNS

In a broad way the kind of soil formed follows the kind of climate in any region of the earth's surface. Within the broad pattern there are variations that are caused by the interaction of the five variables (Section 2–3) in the weathering situation, but the overall effect is one of patterns of pedogenic weathering superimposed on the geochemical weathering of the region.

5–1. Present-Day Soils

In Sections 4–1 and 4–2 the result of pedochemical weathering was shown to be the development of a soil profile with its A, B, C, and D horizons, the D horizon being the parent rock of the soil, and the C horizon being the geochemically weathered D horizon. The common features of weathering in the climatic zones between the north pole and the equator are given in Table 2. In humid regions with abundant rainfall the weathering rocks are leached with water and pedalfer soils result; in dry regions water is scarce, absent, or markedly seasonal and pedocal soils are formed. A number of local variations of the chemical climate cause other kinds of soil to form in the major types of soil.

Soils may be either zonal or azonal. Zonal soils are soils sufficiently mature to have developed profiles that reflect the influence of the soil-forming factors of the climate in which they are situated. Azonal soils are those that have no well developed profiles reflecting the climate in which they are situated. Alluvial soils, sands, lithosols, loess in part, volcanic ash deposits, and some types of soils such as rendzinas on limestone, are azonal. They are too immature to have fully developed profiles; the influence of the parent rock is dominant over climate. In many areas the topography, drainage pattern, and vegetation give rise to a catena of soils, that is, soils developed from the same parent material but in different topographic situations.

In the pedalfer soils from which soluble elements are leached out and alumina and stable minerals remain, the soil forming processes are podzolization, lateritization, and, with impeded drainage, peaty soils and gleization. In the pedocal soils in which lime accumulation occurs through lack of sufficient leaching to remove it, the main soil forming process is calcification. The addition of salts from any source causes salinization in certain areas of pedocal soils.

Podzolization. These soils have a peaty or leafy mat accumulation at the surface, and are highly leached in well-drained conditions. Clay is moved downwards in the profile by a combination of inorganic leaching and chelation with the leachates or organic matter. The soils have an acid reaction with pH around 4.

Lateritization. This is a progressive and accelerated form of podzolization that takes place with time under tropical conditions. Organic matter does not accumulate because of the increased temperature, but it does affect the leaching power of water. At a final stage ferric oxide and alumina accumulate as ferruginous bauxite and laterite. A lateritic soil profile is described later in this chapter.

Peaty Soils. In these soils organic matter accumulates because of waterlogged conditions. Most peaty soils are found in cold climates, but there are also tropical peats, as in Indonesia and Borneo.

Gleization. Imperfectly drained soils with a gray or bluish color due to ferrous iron are known as gleys. Such soils generally have well developed horizons and a mottled zone. The gley is the lowest part of the profile where oxygen is excluded and ferric iron is reduced to ferrous iron.

Calcification. Redistribution of $CaCO_3$ and $MgCO_3$ in a profile without removal results in calcification. Sometimes the lime is finely divided and almost unrecognizable; elsewhere the lime occurs in distinct concretions. These soils have dominantly grassy vegetation. The carbonates give base saturation to the clays and produce a granular condition.

Salinization. Salts accumulate in the soil: NaCl, Na_2SO_4, $NaHCO_3$, Na_2CO_3, $CaSO_4$, and $CaCl_2$ (and sometimes $MgSO_4$ and $MgCl_2$ or K^+ salts). These are also known as alkali soils, and for all salts except NaCl these soils occur as patches in poorly drained parts of other soils. Agriculturally they are known as "white alkali," containing all salts except Na_2CO_3, and K_2CO_3 and "black alkali," containing Na_2CO_3 and K_2CO_3;

the latter cause dispersion of organic matter in a finely divided state and hence the black color.

C. F. Marbut, a pioneer soil scientist in the U.S. Department of Agriculture, observed the climatic soil patterns and divided soils into great soil groups following the Russian soil investigators. The geographic distribution of these great soil groups has formed the basis of soil science throughout the world. In particular, Marbut's ideas have influenced the course of soil mapping and evaluation in the United States, Australia, and New Zealand, and other countries to a lesser extent. These great soil groups are given in Table 8.

5–2. Classification of Soils

Although soil has been recognized as such from the time that man first engaged in agriculture, and all the great cultural systems have been built on agriculture (for example, the Egyptian culture was based on the Nile alluvium), it is only since the beginning of the 19th century that a science of the soil has been developed. The practice of agriculture had been highly developed since before the Christian Era; many of the simple techniques are very old indeed and their origins lost in antiquity. Soil maps of one kind or another have been available for a very long time. The Chinese made a schematic soil map of their country about 42 centuries ago as a basis for taxation and the administration of agricultural matters. Western culture forms the basis of agriculture in Europe, North America, Australia, and New Zealand, but soil science received a tremendous stimulus from Russia which remained a medieval empire controlled by a landed aristocracy (Kellogg, 1938) during the periods of social upheaval in Europe. Soil science developed in Russia along entirely different lines from those in Europe. In the vast area of the Russian Platform, the close relationship between climate, vegetation, and soils was noted. In about 1870 a unique school of soil science was initiated under the leadership of Dokuchaiev, who was followed by Sibertsev, Glinka, Geodroiz, and others. These were all men of vision. In contrast to the agricultural chemists of western Europe, the Russian scientists first studied soils in the field, and later undertook laboratory examinations of the soil materials. They noted that each soil was characterized by a distinct series of layers (horizons) with particular characteristics. Five principal factors were recognized as chiefly contributing to the properties of a soil: climate, vegetation, parent rock, relief, and age (cf. Jenny's factors of soil formation in Chapter 2). Slope and parent rock were more or less passive and of local influence, whereas vegetation and climate were the active factors for regional characteristics and for the great soil belts running through Russia (see Fig. 3 in Chapter 2).

TABLE 8

Great Soil Groups, Marbut's Classification

Soil	Description	Process
Pedalfers		
Podzols	Leaf layer over bleached layer, often with hardpan	Podzolization
Gray-brown podzolic	Thin leaf litter over dark surface soil; leached horizon over heavy B horizon	Podzolization
Red podzolic	Leached surface over deep red B horizon	Podzolization
Yellow podzolic	Thin dark-colored organic layer over pale yellowish gray leached layer; mottled D horizon	Podzolization
Lateritic soils	Brown, friable, leached soils (not to be confused with laterites which have a differently developed profile)	
Pedocals		
Desert soils	Light gray or brown, yellowish, reddish brown, low organic matter	Calcification
Red desert soils	Light reddish brown	Calcification
Sierozem	Pale gray, shallow	Calcification
Brown soils	Brown, calcareous horizon 1–3 ft	Calcification
Reddish brown soils	Deeper soil with heavy subsoil and calcareous horizon	Calcification
Chestnut soils	Dark brown, friable, with lime at 1.5–4.5 ft	Calcification
Reddish chestnut soil	Heavy soil with lime accumulation at 2 ft or more	Calcification
Chernozem soils	Black or dark brown, friable, 3–4 ft deep with lime accumulation	Calcification
Prairie soils	Dark brown to grayish brown	Calcification and weak podzolization
Reddish prairie soils	Dark brown or reddish brown, with heavier subsoil	Calcification and podzolization

From these observations a system of soil classification was made that applied to large areas. Many of the Russian soil terms such as podzol, chernozem, sierozem, solonetz, glei (gley), and others have been adopted universally in soil descriptions. Since the boundaries between climatic belts and vegetative zones were nearly coincident in Russia, much weight was given to the climatic factor, but later investigators have shown that native (endemic) vegetation is more important and that the influence of climate is partly indirect as a determinant of vegetation. The broad classification of soils over large areas gave great impetus to the understanding of soils of similar large areas, such as North America and Australia. Previous ideas of soils had been bounded by the small areas and confines of northern Europe, whose soils largely developed under similar climatic and vegetative regimes. However, colonization of North America and Australia showed that the narrow views of soils resulting from European agricultural practice were untenable in these new areas, and the soil classifications made, although modified, follow the Russian pattern. Within the great soil groups there are individual types of soil profiles that can be recognized and mapped. The unit for profile description is the pedon, which is an area of 1 m^2 to 10 m^2, depending on the variability of the horizons.

Soil maps have been greatly aided by geological maps. In Great Britain, the drift maps issued by the British Geological Survey serve a very useful purpose in soil mapping. In the United States, bedrock geological maps made by the U.S. Geological Survey formed the basis of the soil maps issued by the U.S. Department of Agriculture. In Australia, the tendency was to map soils as units independent of the geology. This arose directly from the influence of Russian soil scientists and from the fact that for vast areas of the country no geological maps were available.

The soils of an area are mapped by soil surveyors in much the same way geologists map rocks. Distribution maps are made of the same kinds of soil profiles, that is, distinct soil types. Some soil maps are very similar to geologic maps, whereas others are very complex. Soil surveyors recognize phase variations of soil types, an eroded phase or a stony phase for example. The phases are the local variations of topography and bedrock within a soil zone. All profiles that are mapped are given names; in the United States this is done by the Department of Agriculture and in Australia, by the Soils Division of the Commonwealth Scientific and Industrial Research Organization. Type descriptions are available for consultation. A broader kind of soil mapping is soil association, a grouping of similar kinds of soil which may form on related rocks, for example, limestones or dolomites that vary only slightly in composition. Soil maps can be used as land-use maps. Such maps show areas of similar inherent fertility and physical characteristics which together influence productivity. Aerial photographs are used to replace much field mapping, thereby saving considerable time.

Large areas, such as North America, Russia, and Australia, serve as models to illustrate pedochemical weathering expressed as soils. In such areas there is a considerable climatic range. In North America the range is from the Arctic (polar) regions to the Gulf of Mexico, that is, between latitudes 80° N and 25° N. In Russia the range is between latitudes 70° N and 40° N, and in Australia between about latitudes 10° S and 45° S. In the distance between north and south in these countries there is a considerable climatic variation between polar and subtropical climates in North America and Russia, and between tropical and cool temperate climates in Australia (Australia being surrounded by water is not subject to the same climatic variations as are North America and Russia).

Marbut's soil classification (Section 5–1) is very useful for describing the major feature of chemical weathering—leaching. It was modified to provide a classification more useful to the description of soils as entities by soil scientists. This classification is given in Table 9.

In 1960 the Soil Survey Staff and the Soil Conservation Service of the U.S. Department of Agriculture published a new soil classification (Soil Classification, A Comprehensive System, 7th Approximation) because of the need to classify soils of tropical, largely undeveloped countries. Prior to World War II modern soil classifications had largely been made by soil scientists from the Northern Hemisphere temperate zone in Europe and North America. These scientists had had little or no experience with tropical soils. For example, there are no laterites in the temperate zone, and only a few types of lateritic soils in the United States. Australia had encountered problems in attempting to classify some of its soils according to the American and Russian standards, and had recognized a soil characteristic of a certain climatic and vegetation region of southern Australia as Mallee soils, which were considered to be worthy of a separate great soil category. In India and South Africa, as well as in Australia, black soils of a somewhat similar superficial appearance to chernozems were recognized, but could not be correctly fitted into the given soil classification scheme.

The soil classification which was suggested in 1960 used new terms to describe soils so that soil scientists from many countries could apply the same terms, which were indicative of appearance and the processes of soil development, to all soils. However, this classification depends on a very good knowledge of the meaning and application of the terms, or the constant use of the book describing the terms. The 1960 soil classification is given in Table 10.

At present the division of soils into pedalfer and pedocal groups (Table 8) is more meaningful when describing weathering than that of

TABLE 9
Classification of Soils Used by the U.S. Department of Agriculture up to 1960

Order	Suborder	Great soil groups
Zonal soils	1. Soils of the cold zone	Tundra soils
	2. Light-colored soils of arid regions	Desert soils Red desert soils Sierozem Brown soils Reddish brown soils
	3. Dark-colored soils of semiarid, subhumid, and humid grasslands	Chestnut soils Reddish chestnut soils Chernozem soils Prairie soils Reddish prairie soils
	4. Soils of the forest–grassland transition	Degraded chernozem Noncalcic brown or shantung brown soils
	5. Light-colored podzolized soils of the timbered regions	Podzol soils Gray wooded or gray podzolic soils* Brown podzolic soils Gray-brown podzolic soils* Red-yellow podzolic soils*
	6. Lateritic soils of forested warm-temperate and tropical regions	Reddish brown lateritic soils* Yellowish brown lateritic soils Laterite soils*
Intrazonal soils	1. Halomorphic (saline and alkali) soils of imperfectly drained arid regions and littoral deposits	Solonchak or saline soils Solonetz soils Soloth soils
	2. Hydromorphic soils of marshes, swamps, seep areas, and flats	Humic gley soils (includes wiesenboden)* Alpine meadow soils Bog soils Half-bog soils Low-humic gley soils* Planosols Groundwater podzol soils Groundwater laterite soils
	3. Calcimorphic soils	Brown forest soils (Braunerde) Rendzina soils
Azonal soils		Lithosols Regosols (includes dry sands) Alluvial soils

* New or recently modified great soil groups.

TABLE 10

Present Soil Orders and Approximate Equivalents in Revised Soil Classification, 1960

Present order	Approximate equivalents
1. Entisols	Azonal soils, and some low-humic gley soils
2. Vertisols	Grumusols
3. Inceptisols	Ando, Sol Brun Acide, some brown forest, low-humic gley, and humic gley soils
4. Aridisols	Desert, reddish desert, sierozem, solonchak, some brown and reddish bown soils, and associated solonetz
5. Mollisols	Chestnut, chernozem, brunizem (prairie), rendzinas, some brown, brown forest, and associated solonetz and humic gley soils
6. Spodosol	Podzols, brown podzolic soils, and groundwater podzols
7. Alfisols	Gray-brown podzolic, gray wooded soils, noncalcic brown soils, degraded chernozem, and associated planosols and some half-bog soils
8. Ultisols	Red-yellow podzolic soils, reddish brown lateritic soils of the U. S., and associated planosols and half-bog soils
9. Oxisols	Laterite soils, latosols
10. Histosols	Bog soils

the 1960 classification (Table 10). The great-soil-group category in soil classification gives the main characteristics of the pedogenic factors that form the soils of any region that has a recognizable steady-state chemical environment. Within the great soil groups families of soils can be recognized as having certain characteristics in common, e.g., families in the low-humic latosols of the Hawaiian Islands. These families are, Molakai, Lahaina, Wahiawa, Kohana, Kohala, Waialua, and Waimanolo, each of which has a recognizable soil profile. Within the family, series of soils occur. The classification of these soils is given in Table 11.

5–3. Soils of the United States

The pedochemical weathering that has taken place in the nonglaciated coterminous United States serves as an example of soil formation over approximately three million square miles of land under various climates and vegetative cover. The U.S. Department of Agriculture has issued a map of the soil association recognized in this area. Figure 10 has been modified from this soil map to show the regional soil patterns that have developed.

It can be seen in Fig. 10 that the western third of the United States has desert soils and lithosols, that there is a belt in the center ranging from

TABLE 11

Classification of Low-Humic Latosols

Great soil group	Family	Series*	Remarks
Low-humic latosol	Lahaina	Lahaina silty clay	Brown to dark reddish brown; uniform in texture and chemical composition; silty clays or silty clay loams with kaolinite as the clay mineral; develops under annual rainfall of 20–40 inches; sea level to 1500 ft; cultivated for pineapple and sugar cane

* Nine other series are mappable within this family.

chernozems in the north through prairie soils to chestnut soils and red-yellow podzolic soils in the south, and that in the eastern third of the United States the range is from podzols in the north through gray-brown and brown podzolic soils to red-yellow podzolics in the southeast to podzol-type soils developed in the sands of Florida. Two-thirds of the soils belong to the pedocal group, while the remainder are leached soils of the pedalfer group. Regional characteristics indicate that many features of the soils, such as reaction (pH), leaching or accumulation of chemical elements, exchange capacity, and fertility can be inferred. The basis for grouping these soils is the great soil categories (Section 5–2). Details of these soils follow.

Podzols: developed in cool, temperate, humid climates, particularly on sandy bedrock; podzols occur in Vermont, New Hampshire, western Massachusetss, Maine, Appalachian plateaus in Virginia, West Virginia, Pennsylvania, southern New York, near Lake Superior, the sandy Coastal Plain from southern Delaware to Cape Cod, and in northcentral Minnesota.

Rainfall: 25–50 inches per year; often snow in winter; 100–150 frost-free days per year; humid.

Parent materials: generally highly siliceous; glacial till and drift derived from shales, sandstones, schist, granite, gneiss, unconsolidated sands and gravels, occasionally limestone.

Brown Podzolic Soils: imperfectly developed podzols with an organic mat at the surface and a very thin leached horizon below it; a small group of soils south of the main podzol belt in New England, parts of New York, New Jersey, and the Connecticut Valley.

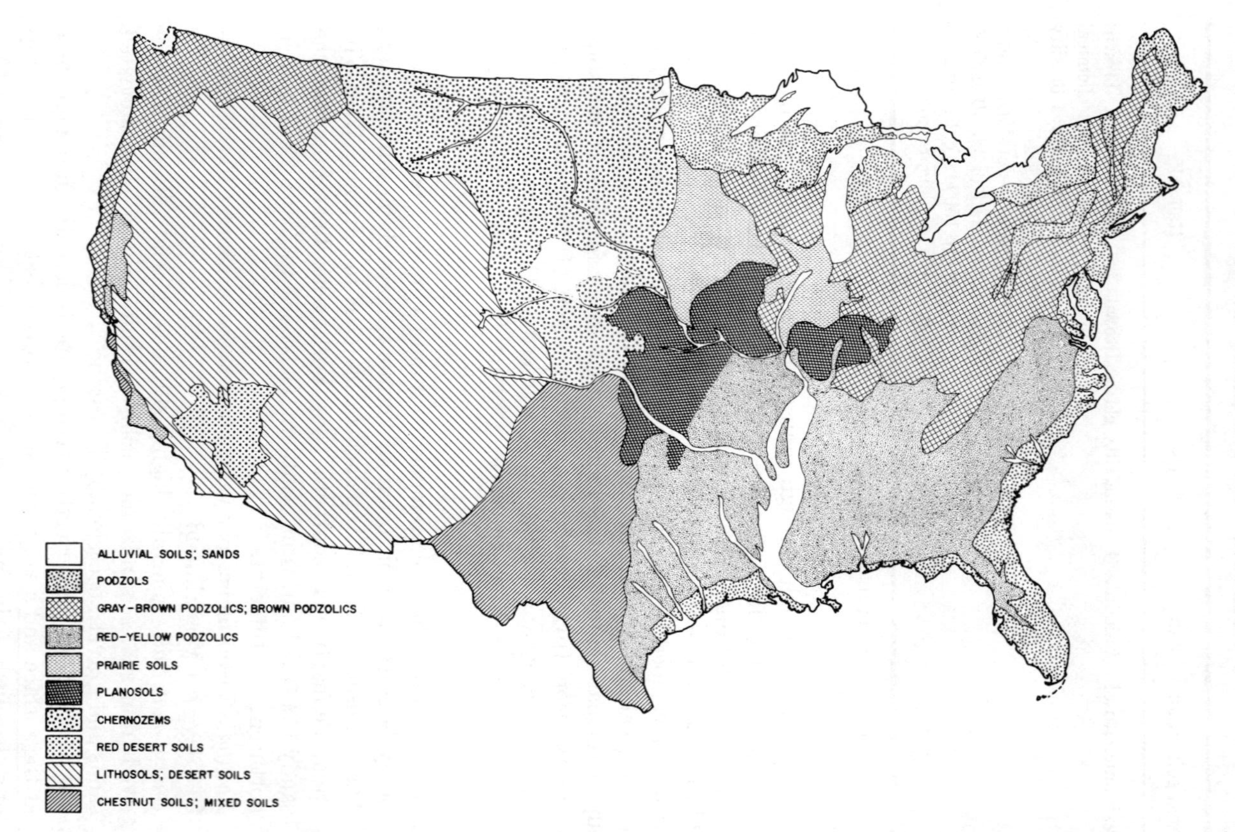

Fig. 10. Generalized soil map of the United States (modified by the author from the U. S. Department of Agriculture's soil association map).

Rainfall: generally 40–50 inches annually: humid; 140–180 frost-free days per year.

Parent material: glacial till and drift from granite, gneiss, schist, slate, Triassic sandstones and shales.

Gray-Brown Podzolic Soils: developed in the eastern and midwestern United States under deciduous forests in a humid, temperate climate; these soils are situated between the podzols in the north and the red-yellow podzolic soils in the south and adjoin the prairie soils in the west; they are a diversified group with a wide distribution; important types are the Chester soils of the Piedmont region in Pennsylvania, New Jersey, Delaware, Maryland, and Virginia; the Fairmont soils of the outer blue-grass area; Hagerstown and Frederick soils in the Appalachian Valley of Maryland, Virginia, Pennsylvania, and northeastern Tennessee; the Miami soils of the little corn belt in Indiana and southwestern Ohio; the Muskingum soils of the high country of West Virginia, Virginia, Pennsylvania, and southeastern Ohio; the Sassafras and Collington soils on the greensands of the Coastal Plain of Maryland, New Jersey, and Virginia.

Rainfall: 25 inches and over annually.

Parent materials: loess, glacial drift, gneiss, granite, basalt, rhyolite, sandstone, limestone, greensand.

Prairie Soils: developed under cool, moderately humid climates that support a grass vegetation in the Middle West; prairie soils cover a large part of the corn-belt country; the soil profiles are characterized by a dark brown to black, mildly acid surface underlain by brown well-oxidized subsoils; the principal areas where these soils are developed are in Kansas, northern Oklahoma, Missouri, and Nebraska; there are similar soils in parts of California, northern Idaho, northeastern Oregon, and southeastern Washington.

Rainfall: 25–30 inches annually with 140–160 frost-free days per year.

Parent materials: calcareous glacial drift, limestone, calcareous shale, loess of Peorian age; the soil on loess may have had calcium carbonate leached away to a depth of 20 inches as a result of the soil-forming conditions persisting for considerable geologic time.

Reddish Prairie Soils: developed south of the prairie soils in south-central Kansas, Oklahoma and north-central Texas in a warm, temperate climate.

Rainfall: 28–35 inches annually.

Parent materials: red calcareous clay or sandy clay, limestone, sandstone, gypsum-bearing sediments.

Red-Yellow Podzolic Soils: these soils are lateritic, i.e., strongly leached, acid in reaction, and low in organic matter; the surface is light colored and sandy, and the subsoil is heavier, tougher, and red, yellow, or mottled; the climate under which they develop is warm, temperate to subtropical and humid; red-yellow podzolic soils occur in the southeastern United States on the Coastal Plain, much of the Piedmont, Ozark plateaus, and in the limestone valleys of the southern part of the Appalachian Plateau.

Rainfall: 40–50 inches annually; 150–300+ frost-free days per year.

Parent materials: these vary in different areas; the Piedmont soils are on granite, gneiss, schist; in the slate belt they are on Carolina slates and basic igneous rocks; in the coastal plain they are on sandy clay, sands, and in the gulf states on clays; in the Valley and Ridge province they are on limestone, sandstone, shale; and in the Mississippi River area, on loess and stream-deposited materials.

Chernozem Soils: sometimes called black earths, and allied to the black soils in India, Africa, and Australia; chernozems are developed in temperate, subtropical grasslands in a broad belt in the central part of the United States, Kansas, south-central Nebraska, South Dakota; the surface soils are dark brown to black; calcium carbonate accumulates at depth.

Rainfall: 18–30 inches annually.

Parent materials: Cretaceous shales (Pierre Shale), Tertiary calcareous sandstones, calcareous glacial drift, lacustrine deposits, Peorian loess, Palouse loess.

Chestnut Soils: these soils have a dark grayish brown surface grading into light gray or white calcareous horizons at about 18–24 inches or more; they occur in the northern Great Plains west of the chernozem belt, parts of Washington, Oregon, and Idaho.

Rainfall: chestnut soils are most frequently developed under 14–18 inches of annual rainfall.

Parent materials: alluvial outwash, loess, calcareous sandstone and shales (Tertiary), calcareous glacial till.

Reddish Chestnut Soils: developed on grassy plains from southern Kansas to Texas and the Gulf of Mexico under a warm, temperate and semiarid to subhumid climate; the surface soil is friable, dark reddish brown-the subsoil is heavier and tougher, reddish brown to red in the upper part, and lighter, or gray, and calcareous in the lower part.

Rainfall: 15–30 inches annually; 175–290 frost-free days per year.

Parent materials: marly and sandy Quaternary deposits, Coastal Plain (Gulf) sands and clays, clays, shales, limestones (Ogallala formation).

Brown Soils: developed in the western part of the Great Plains and in smaller areas in the intermountain region of the Far West; the surface is brown which grades at 1–2 ft into light gray or white calcareous layers; temporate or cool, temperate, semiarid climate.

Rainfall: irregular, 10–20 inches annually.

Parent materials: marl (Ogallala formation), clays, shales, sandstones, glacial drift, loess, colluvium.

Reddish Brown Soils: occur in semiarid areas of the southwest from western Texas to southern Arizona under a warm, temperate climate with hot summers; the surface soil is reddish brown to red, and the subsoil ranges from a red heavy soil to pink and almost white and calcareous at depth.

Rainfall: 10–20 inches annually, with 280–300 frost-free days per year.

Parent materials: Coastal Plain (Gulf) formations, many of which are calcareous, calcareous clay, sandy clay, alluvial material with small included areas of bedrock residuum of granite, syenite, and other acid igneous rocks.

Noncalcic Brown (Shantung) Soils: occur in mountains, hills, or intermountain valleys of southern and central California and central Arizona; these soils are developed under a warm, semiarid or subhumid climate with a cool, moist winter and hot, dry summer; the surface soil is brown, reddish brown, or red, mellow or somewhat compact; the subsoil is heavier, tougher, and redder; it is commonly leached of calcium carbonate, but the pH is neutral or slightly alkaline.

Rainfall: 10–30 inches annually with 210–240 frost-free days per year.

Parent materials: old alluvial fan and valley materials derived from granitic rocks (granite, gneiss, granodiorite).

Sierozem and Desert Soils: occur in desert areas in the western intermountain region under an arid, warm to cool, temperate climate; the surface soil is light grayish brown or gray and low in organic matter, and the subsoil is lighter in color and contains much calcium carbonate.

Rainfall: 3–12 inches annually.

Parent materials: sandstones, shales, limestones, alluvial outwash of these rocks, old lake sediments, sandy alluvial terraces.

Red Desert Soils: occur in the hot, arid southwest from Texas to southeastern California; the surface soil is light colored, pinkish gray, reddish, brown, or red, with reddish brown or red heavier and more compact upper subsoil, and pink or white very calcareous lower subsoil.

Rainfall: 3–12 inches annually.

Parent materials: alluvial outwash from a variety of rocks, residuum of granite, basalt, sedimentary rocks.

Planosols: have developed on flat upland areas of restricted drainage in the Middle West, and in parts of Texas and California; these soils have a

well-defined accumulation of clay or cemented material at varying depths below the surface.
Rainfall: 10–45 inches annually.
Parent materials: loess, glacial drift, stream terrace and alluvial fan deposits of mixed materials, calcareous clays and marls.

There are a number of intrazonal types of soil that occur in various geographic and topographic situations within the areas covered with zonal types. These soils may be caused by microclimates, by influence of parent rock, by accession of atmospheric materials, by recycling of weathered products, and other factors. Such soils include the following:

Rendzina: an intrazonal soil with dark gray or black surface overlying soft, light gray or white, highly calcareous material; these soils are regarded as immature as the influence of the parent material causes the soil type; rendzinas are found in the coastal regions of California, Blackland and Grand Prairies of northeastern Texas and south-central Oklahoma, and the black belt of Alabama and Mississippi.
Rainfall: 15–50 inches annually.
Parent materials: calcareous shales and sandstones, chalk, marl, calcareous clays.

Wiesenboden or Groundwater Podzols, and Half-Bog Soils: this is a miscellaneous group of soils that have a high water table and are poorly drained; they occur in the seaward portions of the Coastal Plain from Norfolk, Virginia, to the Altamaha River in Georgia, Texas, northwestern Minnesota, Louisiana, eastern Arkansas, the Great Lakes region, the Maine coast, and the Champlain–Hudson Valley in New York and Vermont.
Rainfall: 30–55 inches annually.
Parent materials: unconsolidated sands, sandy clays, clays, glacial drift, calcareous clays and marls.

Solonchak and Solonetz Soils: salty and alkaline soils of the arid and semiarid regions of the western United States; they are either poorly drained or developed under poor drainage conditions; solonchak soils contain large quantities of soluble salts, are light-colored, poor in organic matter, and have a lightly crusted, friable, granular structure; solonetz soils are partially leached and alkalized solonchak soils; they have a thin surface layer of light-colored, leached material over a darker colored subsoil of tough, heavy material with a columnar structure; the lower subsoil is light gray and highly calcareous; such soils occur in the dry interior valleys of California (e.g., the San Joaquin Valley), and the Great Basin in Utah and Nevada.
Rainfall: 3–15 inches annually.

Parent materials: alluvial and lacustrine deposits of sandy or clayey material.

Lithosols and Shallow Soils: these are immature, thin soils or residual material that occur on various rock formations, and undifferentiated rough and stony land where physiographic conditions or time have not been favorable for the development of mature soils; lithosols are found in regions of active erosion.

Bog Soils: include peat and muck, and soils in coastal marshlands; peat and muck have an area of about 79 million acres (123,450 square miles) in the United States with extensive areas in the southeast, the Great Lakes region, and the Pacific Coast; the Everglades in Florida are typical of these soils; coastal marshlands occur in many areas along the Atlantic and Gulf coasts.

Alluvial Soils and Dune Sands: see Sections 5–9 and 5–8.

5–4. Lateritic Soils, Laterite, and Bauxite

Lateritic Soils

The podzolization process is one of leaching that removes all the soluble material in a soil profile. Podzols, as noted earlier, develop in cool, moist climates under a leaf cover, preferably of pine needles. The water passing through the raw humus mat contains chelating substances (mainly fulvic acid, (Section 9–3), the effect of which is to increase the removal of iron and aluminum from the developing profile. When a strongly leaching condition occurs in a tropical climate, lateritization takes place. The humid tropics provide the most intense conditions under which soils develop, and of the several types of tropical climates, that which has no dry season produces the greatest amount of weathering. This climate would be classified as Afa in Köppen's classification (Fig. 3, Section 2–2). Such climates are found in Malaya, Nigeria, northeastern South America, and in some of the Hawaiian Islands (Kauai). Under these conditions, after the more soluble constituents of geochemical weathering are leached out, silica is removed, and, consequently, alumina, ferric oxide, and titania accumulate. The soils subjected to this type of leaching are acid in reaction and have little or no plant-food elements. Some resistant or stable minerals may remain and accumulate. Such soils are unattractive for agriculture and can only be used for small subsistence farming and for certain crops. The lateritization process explains why it is so difficult to develop agriculture in tropical countries.

Laterite

This name was originally given to a soft, weathered rock used in India and other southeastern Asia countries for building. The material, largely a moist saprolite, could be cut into blocks with a knife, and when the blocks dried they became quite hard and weather-resistant. Although laterite had been used for many centuries for building, the first description of laterite was made in 1807. A description of the usage of this material is given by Prescott and Pendleton (1952). Early in the 20th century, Australian geologists recognized the uppermost zones of the lateritic soil that occurred over wide stretches of southwestern Australia as fossil laterite, similar to the Asian laterites. These fossil laterites had probably been developed in the Pliocene geologic period, and because of lack of erosion are preserved as remnants of a once very extensive weathering that covered hundreds of square miles.

A section through laterite from the surface to the underlying rock generally shows different horizons, and is, in fact, a soil profile:

Surface: pisolitic and enriched in iron oxides; sometimes there is a sandy layer above the pisolites, particularly if the profile has a granitic parent rock and severe weathering conditions have continued for considerable geologic time; the sand-plains in southwestern Australia are an example.

Mottled zone: generally white clay with red mottles or vertical accumulations of iron oxide.

Pallid zone: a bleached zone between the mottled zone and the saprolite beneath.

Parent material: the pallid zone grades down into a highly weathered rock or saprolite and then into fresh rock.

As has been noted in Hawaii and elsewhere, lateritic profiles do not always have mottled and pallid zones. A laterite profile may be 50 or more feet deep. Figure 11 shows the sequence of layers in a laterite developed from a granitic parent rock. There are stages other than the fully developed stage in the production of a laterite that can be recognized. These stages are produced by lateritizing processes which cause the gradual leaching out of silica and the accumulation of alumina and iron. The red-yellow podzolic soils are examples of the lateritic process, and although no true laterites occur in the coterminous United States, the soils of the Tifton–Irvington area in southwestern Georgia are examples of lateritic soils.

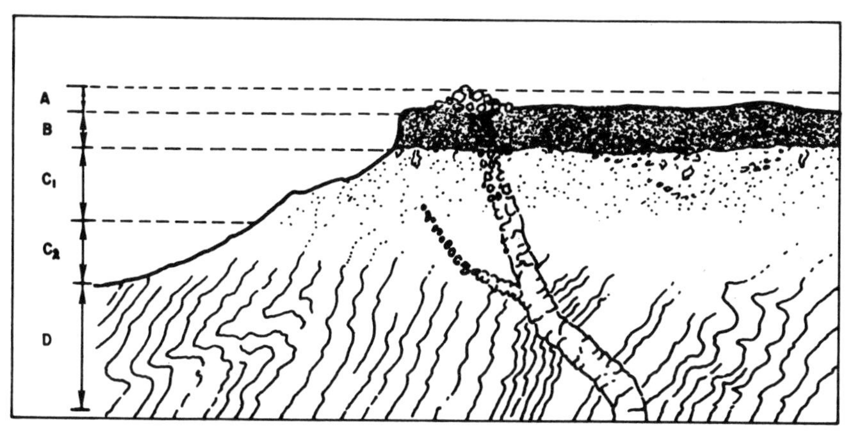

Fig. 11. Lateritic soil profile developed on saprolitized Precambrian gneiss. This illustration is modified from the original published in 1915 by Walther, who recognized the erosion feature, known as a "breakaway," in the interior of Western Australia. A) A horizon, generally sand, largely removed by erosion (see Section 5-7). B) B horizon, concretionary laterite. C_1) Mottled zone. C_2) Pallid zone. D) Saprolite with a quartz vein.

The stages in the production of a laterite from an acid igneous rock are:

1. Initial weathering of rock	Silica is partly mobilized; feldspar surfaces react with percolating water and some cations are removed from feldspars and ferromagnesians.
2. Production of saprolite	Feldspars alter to clay minerals of kaolinitic and micaceous type; ferromagnesians lose FeO by oxidation to Fe_2O_3; olivine (not present in granites) alters to montmorillonite; nepheline alters to gibbsite; texture and structure of the rock are not changed, although the original minerals have been replaced by alteration products.
3. Saprolite at the surface is strongly leached and a soil starts to form	Stages 1 and 2 proceed well below the ground surface and are geochemical weathering; in the soil, silica removal continues, but quartz accumulates because of its insolubility; it is the combined silica that is removed (pedochemical weathering); ferric iron is concentrated into a horizon similar to that in podzols.
4. Waterlogged conditions	Gleying occurs at depth in the profile and iron is mobilized as $Fe(OH)_2$ when air is excluded; mottles develop and the pallid zone is formed.
5. Intermittently dry conditions	Formation of pisolites in "podzolic" B horizon; accumulation of quartz and stable heavy accessory minerals at the surface of the profiles developed on granites and other siliceous rocks.

6. Dry conditions — The silica dissolved from feldspars and other weatherable minerals accumulates at the top of the water table, dries out to a gel if drainage is poor, and surrounds other grains to form a quartzite-like material; it is known as "billy" in Australia; Fig. 12. is a photomicrograph of this material.

In thin section the siliceous rock, or billy, of Fig. 12 has much the appearance of a sandstone with the individual quartz grains floating in a matrix of opaline silica. The quartz grains are angular to well rounded in shape without any secondary outgrowths. Billy apparently results from the removal of silica from the minerals of the parent rocks of laterite profiles and probably is an expression of the difference in solubility of quartz and the silica combined in minerals (see Fig. 21 in Section 8–3). Much of the dissolved silica would be in an amorphous state and metastably in solution at about 140 ppm (Morey, Fournier, and Rowe, 1964). Because this silica was not removed by drainage due to the flat topography and lack of rivers, it remained with the other weathering products of laterite profile development. In Australia, with very extensive plateau areas and poor drainage, siliceous rocks like billy are common. Mohr and van Baren (1954; pp. 380–387) have summarized the pioneer observations of Whitehouse in Queensland concerning silication of weathering products, and in Western Australia Teakle (1937) pointed out that there are some 200,000 square miles of soils in the arid interior with siliceous red and brown hard-pans which must have formed as a result of transportation and consolidation of amorphous silica during laterization of gneisses of the western part of the

Fig. 12. Photomicrograph of silcrete, known as "billy"; South Australia. Scale is 1 mm. This section, crossed nicols (photograph by N. Prime).

Australian Precambrian shield. Alexander (1959) made some very pertinent observations about the movement of silica and alumina under intense tropical weathering conditions. Such conditions are probably similar to those under which laterite was developed in Australia.

The process of lateritization is similar in all kinds of rocks, but there is a larger quantity of removable silica in basic rocks than in granite, because in the latter much of the silica is in the form of quartz and is relatively inert. The variation in mineralogy between granite and basalt (as end members of a series) is:

	Granite	*Basalt*
Minerals present:	Quartz, feldspars, micas, ± hornblende	Feldspars, pyroxenes, ± olivine, iron ores
Constituents:	SiO_2, about half in quartz, remainder in other minerals; Fe very little, in micas, hornblende	SiO_2, all combined in feldspars, pyroxenes, olivine; Fe, much, both in minerals and crystallized as magnetite and ilmenite;
	Al_2O_3, in feldspars and micas	Al_2O_3, in ferromagnesians and feldspars
	TiO_2, only a little in micas	TiO_2, often much, in pyroxenes and ilmenite (relatively inert under oxidizing conditions)

The minerals that are found in laterites developed from each of the above parent rocks are the same, but vary in quantity. These minerals are kaolinite, gibbsite (sometimes boehmite and diaspore), goethite and/or hematite, and leucozene and/or anatase. There may be small quantities of other minerals, but in granitic laterites quartz is the important constituent. The end product for a basaltic laterite will be one in which all the SiO_2 and Fe_2O_3 have been removed, leaving gibbsite and some form of TiO_2, both of which are impossible to remove by leaching. The same stage in a granitic laterite will be a mixture of kaolinite, gibbsite, and quartz, the latter remaining almost inert throughout, but acting physically as a mesh structure (sieve) which facilitates the movement of water containing other constituents. In other words, quartz grains facilitate drainage. If for any reason, such as the permanent alteration of the water table level, the leaching process should cease or become less efficient, SiO_2, which had reached a certain level in the profile (as a constituent of the leaching water), could accumulate, and on drying out form chalcedony which would persist and inhibit any further development of the profile. Silica might, too, accumulate in slight hollows in a land surface which was being lateritized, this accumulation being due to lateral transfer of mobilized SiO_2 from higher parts of the area (the solubility of silica in feldspars and ferromagnesians may be about 140 ppm, that in quartz is about 6 ppm).

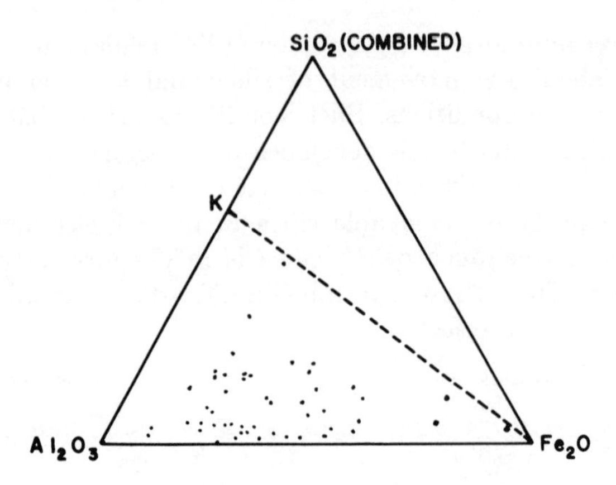

Fig. 13. Chemical composition of lateritic weathering crust in southwestern Australia (after Prescott and Pendleton, 1952).

It is likely that this accumulation of silica will be more marked in laterites developed from basaltic rocks, for there is actually more silica to go into solution from basaltic rocks than from granites. Titania is insoluble under oxidizing conditions, but it becomes somewhat soluble under reducing conditions as Ti^{+4} changes its valence to Ti^{+3}. Under reducing conditions, any rutile grains present go into solution at the edges and anatase and brookite crystallize (Carroll, 1960).

The chemical composition of the very extensive lateritic weathering crust in southwestern Australia is shown in Fig. 13. This diagram was compiled by Prescott and Pendleton (1952) from analyses made by the Western Australian Mines Department. The laterites generally contain a considerable quantity of quartz. The diagram gives figures calculated on a quartz-free basis.

Bauxite

The leaching process that produces lateritic soils and laterites will also produce bauxite from suitable materials. Bauxite is an aluminum ore and consists of the mineral gibbsite ($Al_2O_3 \cdot 3H_2O$) and/or boehmite ($Al_2O_3 \cdot 2H_2O$). There are two principal types of bauxite deposits: blanket deposits that are typical of a tropical, lateritic type of weathering and pocket deposits that occur in limestones. Leaching of SiO_2 leaves Al_2O_3, and the smaller the content of silica the more valuable is the bauxite commercially. Under leaching conditions, gibbsite forms very readily from nepheline and feldspar in syenites. Examples are the Arkansas bauxites and that formed in the Khibny region of the northern USSR inside the Arctic Circle. Many commercial deposites of bauxite are formed in limestones. These are known as the *terra rossa* type, after the soil that is characteristic of weathering in karst topography. Terra rossa soils are common in the Mediterranean region. Examples of this type of bauxite are found in southern France, Hungary, and Jamaica. The bauxite is in pockets and lenses and consists of gibbsite or boehmite. Small pockets of this type also occur in the southeastern United States, particularly in Georgia, Alabama, Tennessee, Mississippi, and Virginia. The bauxites are associated with the Midway Group on the Coastal Plain, and with Paleozoic limestone in the Valley and Ridge province. The deposits are generally of the sink-hole type in which clay minerals have been leached by alkaline solutions from the surrounding limestone.

5–5. Volcanic Ash

In areas influenced by volcanoes of the explosive type in which large quantities of ash are ejected, the mineralogical materials of the ash, when deposited on land surfaces, are an important source of additional material that form part of pedochemical weathering. Such material becomes incorporated in the soils formed. Important factors are the amount of ash deposited and the composition of the ash. The presence of volcanic ash influences the developing soil in mineralogical composition, and hence its weathering features, and in grain-size distribution, or mechanical composition.

It should be noted that most of the ash ejected and deposited is andesitic in composition. Basaltic lava seldom forms ash.

The principal active volcanoes of the present are concentrated in the East Indian Archipelago (Indonesia has over 300) and in a belt running from Japan to New Zealand. Consequently, ash materials are of frequent occurrence in the soils of Java, Sumatra, the Philippines, Papua, New Zealand, and Japan. In Java, in addition to wind-transported volcanic ash, there are mudflows, or lahars, that distribute comminuted volcanic material. The lahars cause a considerable amount of destruction.

Dutch soil scientists have supplied a great many details concerning the admixture of volcanic ash with other soil-forming materials (Mohr and van Baren, 1954). Since some of the earliest work on volcanic ash was done on ash from the eruption of Krakatao in 1883, the following particulars are of interest. Eighteen cubic kilometers of ash was blown into the air and distributed by wind and air currents. The finest material reached heights of 50 km (37.3 miles); when it settled the ash covered an area of 827,000 km^2 (319,220 square miles or an area the size of Texas and

Louisiana). The chemical composition (on a water-free basis) of the Krakatao ash was

	Most basic, %	*Most acid, %*
SiO_2	61.36	68.99
Al_2O_3	17.77	16.07
TiO_2	1.12	0.82
FeO	1.71	1.10
Fe_2O_3	4.39	2.63
CaO	3.43	3.16
MgO	2.32	1.08
MnO	0.41	0.28
K_2O	2.51	1.83
Na_2O	4.98	4.04

and consisted of

Pumice fragments	70.00%	
Glass fragments	21.00	91% glass
Plagioclase (oligoclase–andesine)	6.00	
Hypersthene	1.36	7% crystallized minerals
Augite	0.64	
Magnetite	1.0	

Other volcanic ashes are similar but vary slightly in percentage composition.

In 1931 J. van Baren examined the soil which had been formed in 50 years from the pumice of Long Island of the Krakatoa group. He found no indication of a difference in mineralogical composition between ash and soil, and there was no weathering effect observable on the individual mineral grains. Andesine was the dominant constituent with associated green hornblende, augite, and hypersthene, representing a hypersthene andesitic magma.

In connection with the agricultural development of the land, Dutch soil scientists found that determination of the minerals in these ash-contaminated soils gave a clear indication of the reserve of chemical elements that would become available gradually as these soils weathered. It should be remembered that most soils of the tropics are leached of their soluble constituents by the lateritizing process that is the common method of weathering under tropical conditions.

The ash from other active volcanoes was deposited at varying distances from the volcanoes. In addition to these historically recent ash falls, eruptions also occurred during the Tertiary, but the mineralogy of all these falls, even from the same volcano, may not be similar. For example, Mohr and van Baren (1954) describe the ash of the Gunung Salak near Bogor, Java. The andesitic ash of this volcano weathered into a yellowish brown lateritic clay, but beneath this soil, at about 1.5 m below the surface, there is a light yellow sandy layer consisting largely of coarse crystals of plagioclase. This layer has a thickness of 25 cm at Bogor. It wedged out to the north at a distance of 25 km, but in the southwest, on the volcano slopes, it is 1–1.5 m thick. A cut through such weathering material shows its composite nature. The amount of ash from a volcano always varies with distance from the volcano and the wind conditions. Thick beds are readily recognizable, particularly if the ash is unweathered.

In addition to the mineral constituents of an ash, there may be gases, such as sulfuric and hydrochloric acid, that have an effect on the course of weathering and may be exceedingly harmful to vegetation. One interesting feature is that the emanation of volcanic gases may cause, as it does in Hawaii, an acidity in the rainfall. The effect is that the rain is more acid (lower pH) than it should be in islands surrounded by oceans (Carroll, 1962).

In many parts of the world where volcanic ash has been incorporated during the course of pedochemical weathering, the presence of certain minerals such as augite, hypersthene, and magnetite enable such ash to be recognized. In some areas volcanic ash has been contemporaneously mixed with chemically precipitated limestones (Hathaway and Carroll, 1964).

In the United States, beds of volcanic ash have long been recognized as key horizons in various formations, particularly the Cretaceous and parts of the Paleozoic where the ashes have been metamorphosed to beds of metabentonite. The extensive bentonite beds in Wyoming and North Dakota originated as altered volcanic ash, now montmorillonite. Unconsolidated ash deposits occur in the Pliocene and Pleistocene of western Kansas (Swineford and Frye, 1946). When rocks that contain either dispersed or banded volcanic ash weather, the constituent minerals have a marked influence on the mineralogy of the resulting soil. The ash beds in part of the Niobrara limestone, for example, increase the content of montmorillonite in the soil. A similar situation exists with the weathering products of the Pierre Shale.

5–6. Loess

Loess is an eolian, unstratified deposit of silty mineralogical material that ranges in size from the finest clay to fine sand, but predominantely silt, and has been deposited over extensive areas of North America, northern Europe, and Asia. The name, loess, comes from the German *löss* (from *losen*—to dissolve, pour, or loosen). Loess is yellowish brown in color and is generally calcareous; it may contain concretions of calcium carbonate of various shapes, the more elongated of which are known as *Löss-püppchen* in northern Europe. Loess originated as silt in periglacial areas that were associated with Pleistocene glaciations. The upland silt in Alaska (Péwé, 1955) is similar material that forms a kind of loess at the present time. As loess everywhere originates in the same way and is transported by wind it has the same-sized material, but mineralogically it reflects the composition of the rocks from which the silt was derived, though wind action has a winnowing effect on the original composition of the silt.

One of the best known and most extensive loess occurrences in the world (Leighton and Willman, 1950) is in the Mississippi River basin, comprising over 200,000 square miles in the states of Minnesota, Wisconsin, Illinois, Iowa, Nebraska, Missouri, Kansas, Arkansas, Ohio, and Mississippi. The loess deposits occur in stratigraphic succession and may be interbedded with drift sheets or upon the surface of drift sheets. Loess was deposited by wind action on valley trains of fine glacial outwash that accumulated in silt-covered flats. When glacial floods subsided in the fall the silts were exposed, dried, and the loose surface material was continually removed by winds that swept down the valleys. The wider the valley, the greater the area over which the silt spread, and therefore the greater the amount which could be removed by wind. The climate during the time that loess was being deposited varied from periglacial near the ice front to temperate farther south, though somewhat cooler than the present (Leighton and Willman, 1950, p. 622). The winds were mainly westerly, but changed to northwest during the fall when the maximum loess deposits were made. The faunas found in the loess indicate that rainfall and temperature were adequate to support a forest-type vegetation near the valleys, and prairie vegetation on the flat uplands.

Loess deposits in this region vary in thickness. They form bluffs adjacent to the Mississippi and its major tributaries that may be 60 or more feet high, but the thickness decreases with distance away from the river (Smith, 1942). The coarser grains in the loess decreases in quantity and the finer increase with distance from the river bluffs. However, the loess is homogenous as to size. Figure 14 shows the grain size of loess from the Peorian (Tazewell), Farmdale, and Loveland loesses of Illinois and Ohio.

The loess of the Mississippi River basin is calcareous, and after deposition calcium carbonate was leached out as pedochemical weathering began. The amount of leaching is commonly greater away from the main river bluffs owing to differences in thickness of the loess and consequent greater porosity of the thinner loess deposits. Smith (1942) calculated the average loss of $CaCO_3$ from the Peorian loess to be about equal to an 18-inch layer of loess. The leaching loss during the entire period of depo-

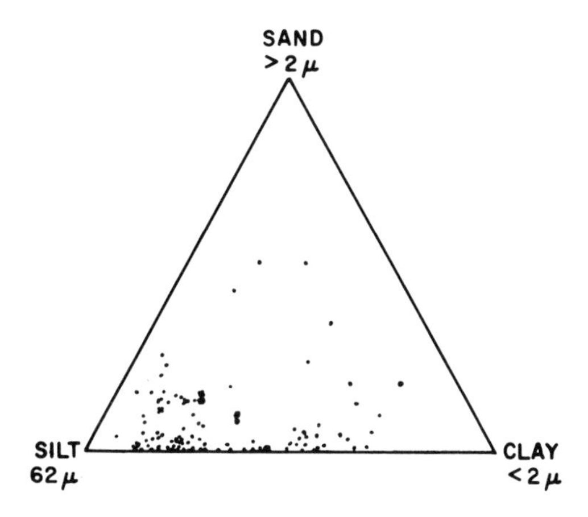

Fig. 14. Grain size of loess from Illinois and Ohio, small dots (after Paul D. Blackmon, unpublished), and from Europe, large dots (after Swineford and Frye, 1955).

sition was calculated to be equivalent to about a 30-inch layer of $CaCO_3$. The period of deposition was therefore a much longer period than the subsequent weathering period.

Very extensive loess-covered areas occur in southeastern Washington, northeastern Oregon, and northern Idaho, and along the Columbia and Fraser rivers in British Columbia. Loess covers the surface of the latest flow of basalt of the Columbia River Plateau in various thickness from an inch or more to about 25–30 ft near the Columbia River. The loess provides a soil of varying thickness over the practically unweathered basalt and forms fertile agricultural land in a region which has developed on country rock. The well known Palouse loess occurs in the bend of the Columbia River near Pasco and Kenniwick.

Loess from a number of type localities in western Europe was collected by John C. Frye in 1952 and examined and compared with Kansas loess (Swineford and Frye, 1955). The European loess is similar in grain-size distribution to the North American and Dutch loess (Doeglas, 1949). The buried soils (Section 11–3) from Europe are similar to those of the Great Plains region, but various other features of the European loesses are not always similar to those of North America. The clay minerals in the European loesses are a 14A chlorite–vermiculite and a well-crystallized mica, whereas montmorillonite is common to the Kansas loess and is accompanied by illite. Thus, although the field occurrence of loess in western Europe and North America is similar there is a considerable variation in mineralogy which probably reflects the dissimilarity of source areas. The montmorillonite in the loess from Kansas may be due to an admixture of volcanic glass, as montmorillonite is not abundant in the glacial deposits of North America, nor is it likely to have come from the source rocks comminuted by glacial action and largely unaltered mineralogically.

The loess of the Netherlands is also eolian in origin, and over extensive areas the grain-size distribution is similar to that in most areas of loess. It is a periglacial deposit from the Wurm (4th northern European glaciation) icecap over Scandinavia. The Riss (3rd northern European glaciation) icecap covered the Netherlands but may be the origin of loess farther to the south. As in North America the loess in the Netherlands carpets hills and valleys and various geological formations. The loess grains range between 2 and 50 μ, but the majority of grains are between 10 and 50 μ (van Doormaal, 1945). The loess was influenced by a cold climate during which some mixing due to solifluction occurred in the Holocene (the western European Holocene includes the earliest part of the Pleistocene to present). Loess was deposited during several periods, and as the covering of the older deposits by the younger was not complete, a mosiac of loess deposits is present. The older loess has been weathered much longer than the younger, and a brown forest soil profile has developed with decalcification of the loess to a depth of over 9 ft. The surface of this type of profile is slightly podzolized and indicates an adjustment of the pedochemical weathering to climate. This weathered loess supported a forest vegetation.

In addition to loess that originated in the periglacial climate (cold loess), there are eolian deposits derived from desert and arid regions (hot loess). Such material was recognized by Butler (1956) in the Riverine Plain of southeastern Australia and called *parna* (Australian aboriginal for "sandy and dusty ground"). Parna has many of the characteristics of loess, but has a greater clay content (30–60+), and the clay is present as aggregates in the size range of 40–400 μ. Parna is the dust-like material that is winnowed from arid regions by winds, as there is little vegetation to prevent wind action. The material of parna forms the dust clouds, similar to clouds of disturbed and blowing loess in exposed situations, in southeastern Australia that produce "red rain." The dust is carried from west to east, or northwest to southeast, by wind, and has been recognized in New Zealand as Australian dust. Like loess, parna forms a mantling deposit in which soil profiles will develop if the rainfall is sufficient to cause chemical weathering. Most of the soils developed are red-brown earths, but in poorly drained situations gray and brown soils of heavy texture are formed. The Southern Hemisphere had no glaciations during the Pleistocene comparable with those in the Northern Hemisphere. However the sheet-like deposits of parna in Australia originated in the climatic fluctuations between arid and humid during the Pleistocene. Loess-like material with a high clay content may be expected to occur at the margins of arid country in Africa and Asia.

5-7. Sands

Sand is an important material on parts of the earth's surface. It is the result of erosion and accumulation. In arid climates sand is formed by the physical disintegration of rocks of all kinds, igneous, metamorphic, and sedimentary. The sand piled up by wind in deserts is largely formed from the disintegration of sedimentary rocks. In moist climates streams remove sand from weathered rocks and from soil profiles, transport it for a certain distance, and deposit it. During transportation the finer material, the silt and clay, is winnowed out of the original disintegrated material and, because of its size, is carried farther from its source than the sand. In the Northern Hemisphere ice ground up the rocks over which it spread at icecaps; the coarser grains of the ground rocks were deposited as outwash sands as the icecaps retreated. Another effect of glaciation was the lowering of sea level which caused sands on the innermost part of the continental shelves to be exposed and transported inland by winds, thus forming dunes.

The complete physical disintegration of a rock produces a sand, but the term usually means a transported material, largely quartz. This definition implies that granitic rocks will produce sand, whereas a basic rock will not, except under desert or glacial conditions. Sand implies a definite size; it grades into grits and gravels on the coarse side, and into silt on the fine side. Sand is sized by being passed through a set of sieves of known, standard-sized openings. There are several different kinds of standards used for describing sands, one of the most satisfactory being the Wentworth scale in which the holes in the sieves are, progressing from coarse to fine, one-half the size of the next larger size. Sand on this scale is between 2 mm and 0.062 mm in grain diameter. Sand means mineral grains of this size—quartz, calcium carbonate, volcanic ash, or any other minerals that occur in this size.

Sand-sized material is generally divided into a number of classes or different sizes within the limits of 2- and 0.062-mm grain diameter; the weight of the various classes compared with the weight of the unsieved sand classifies a sand according to its sizing or sorting by natural agents, e.g., wind or water. A crushed rock made from a piece of fresh rock will have a completely random grain frequency, which, if made into a graph using percent weight of each grade cumulatively on the ordinate and the grades as abscissa, will give a normal Gaussian distribution curve (Krumbein and Pettijohn, 1938). If the sand has been size-sorted some grades

TABLE 12

Grain size, mm	Grade, %						
	A	B	C	D	E	F	G
+2.0	—	—	0.4	—	1.65	—	5.7
+1.0	0.03	—	0.9	—	8.08	—	3.7
+0.5	19.52	—	12.6	0.1	26.05	1.4	6.9
+0.25	61.07	14.0	62.7	70.6	18.27	93.5	26.7
+0.125	13.75	76.4	22.8	26.6	32.24	5.1	34.2
+0.062	4.58	9.3	0.4	2.8	12.51	—	14.8

A) sand from Egyptian Desert; B) sand from Simpson Desert, Australia; C) coral-beach sand, Taongi Atoll, Marshall Islands; D) glass sand, Fontainblau, France; E) St. Peter sand, derived from St. Peter sandstone (average, courtesy W. D. Keller); F) dune sand, volcanic glass shards, southern coast island of Hawaii; G) sand-plain sand, Yilgarn Goldfield, Western Australia (a horizon of fossil laterite, Fig. 11, Section 5–4).

will predominate and others may be missing. When a sandstone disintegrates *in situ* the cement holding the grains together is dissolved and the grain-size distribution of the sand will be very much the same as that of the grains in the compact, unweathered rock.

The grain-size distribution or frequency of grades expressed as a percentage of the whole sand for a number of different sands is given in Table 12.

The grain-size analyses in Table 12 show the main features of sands, but in each group of sands there may be variations due to the effect of local conditions on the agent causing the sorting.

Sands are azonal and, because of their composition (resistate), show very few changes when subjected to weathering. They are spread over the surfaces of weathered or unweathered rocks and prevent further development of soils. Glacial-outwash sands and the "deck" sands of the Netherlands have been spread over humid land surfaces, and may in time become weakly podzolized. In addition, the pore spaces of these sands hold water which is released gradually thereby providing a leaching solution for the weathering of the underlying rocks. Dunes in an arid region may become fixed by vegetation if the region, in geologic time, becomes less arid. Such sands cover previously developed soil profiles that may be recognized by profile variation and by mineralogy. Sands of the Maryland Coastal Plain overlap the eastern edge of the Piedmont which consists of weathered Precambrian rocks. Cores through the material show the change in mineralogical characteristics. An example of this overlap is shown in the mineralogical composition of the montalto soil profile of Fig. 8 (Section 4–2). There is a distinct break in sequence at the top of the BC horizon.

5–8. Alluvial Soils

Erosion is a natural geological process that takes place continuously to smooth the earth s surface and to provide the material for new sediments. If the untouched surface of the earth is disturbed by man and his works, such as by crop production, dam building, road-cuts for highways, housing developments, etc., the effect of erosion is increased. It has been stated that within the last century erosion has made cultivation economically unsound on more than 20% of the tilled ground of the United States. Wind and water erosion remove about 3 billion tons of soil yearly from agricultural land in the United States. The Mississippi River system carries 730 million tons of sands, silts, and clays annually into the Gulf of Mexico.

Alluvial soils are the transported eroded A and B horizons of soils produced by pedochemical weathering in humid climates. As the A horizon is removed by erosion, more soil material is generated by the soil-forming processes from the C and D horizons, but the development of a soil is only accomplished in geological time. The first stage in the production of an alluvial soil is the movement of soil material downslope by water (sheet erosion) and gravitational creep into river valleys where it forms a deposit in the flood plain of a stream or river. The material is added to each year. Alluvial soils develop in any topographic situation which has slope to cause downward movement of the material. In humid regions rivers provide the eroding force, and alluvium is found in valleys and river bottoms; in arid regions the colluvium and alluvial material move very slowly downslope in dry valleys. Some desert areas in the world have been formed by the destruction of natural vegetation in marginal areas. Vegetation in dry areas can only be established very slowly because of lack of water, and once destroyed wind erosion moves the silt and clay away from the area and may pile the sand into dunes. It has been estimated that the time required to remove a 7-inch layer of topsoil under a vegetative cover is from about 4000 to 96,000 years, whereas the same soil with a fallowed surface (unvegetated) is removed in 16 to 50 years.

Alluvial soils are the most fertile of all soils because the mineral matter in the soil generally comes from a number of different sources, many of which may be a considerable distance away from the flood plain where the alluvium is deposited, and because the material is well mixed before deposition. The map of the soil patterns of the United States (Fig. 10, Section 5–3) is on too small a scale to show any but the largest areas of alluvial soils; these are principally along the Mississippi River. The alluvial soils of the northeastern United States come from a variety of different rocks, but the areas are small. In the Prairies and eastern Great Plains the alluvial soils come from materials that were originally dark-colored upland soils; these alluvial soils are black and the subsoils are gray or mottled gray and brown and very clayey. The alluvial soils of the arid West occur on bottom lands, flood plains, low terraces and alluvial fans in the western Great Plains and the intermountain region between Mexico and Canada. The largest area of these soils is in the Sacramento and San Joaquin valleys in California.

If land has been uplifted (initiating a new cycle of erosion) a river may cut down through its previous flood plain and alluvium, which will then appear as a terrace. Geologically old terraces tend towards the development of the zonal type of soils compatible with the climatic zone in which they occur. The south fork of the Shenandoah River flows in an alluvial plain of this type, and there are many other examples in other countries.

Chapter 6

AMOUNT OF CHEMICAL WEATHERING

Direct comparison of chemical analyses of fresh and weathered rocks shows that some constituents have been lost from fresh rock during weathering, and that some have accumulated in the weathered material. Some minerals are more stable than others to the weathering processes, and some chemical elements react with the watery solutions that cause alteration, whereas others do not.

6–1. Calculating the Amount of Chemical Weathering

Weathering in humid climates is primarily caused by water and oxidation, and the several ways in which the amount of chemical weathering can be calculated are measures of these chemical processes. A direct comparison of the chemical analyses of fresh rocks and the soils developed from them yields a considerable amount of information about the chemical processes involved. By using an index mineral, one that is common both to the rock and to the soil and is considered to be unaltered by weathering, one can calculate how much rock has weathered to produce the surface soil. The normative composition of a soil can be calculated from a chemical analysis, and from a comparison of the water added the change in chemical composition between rock and soil can be calculated. By assuming a "standard cell" of 160 oxygen ions, Barth (1948) showed how chemical weathering could be measured. The amount of clay in a soil developed *in situ* above a rock containing weatherable minerals is also a measure of weathering.

6–2. Barth's Calculations

Barth has shown that if the composition of a standard cell of 160 oxygen ions of a fresh rock is compared with that of a weathered rock,

TABLE 13

Rock or mineral	Number of cations	Number of oxygen ions
Average igneous rock	99.3	160
Average basalt	102.4	160
Granite	96.2	160
Quartz	80	160
Kaolinite	71	160

the amount of alteration is easily seen. In most rocks oxygen makes up 92% of the volume, and all cations present (Si, Al, Fe, Mg, K, Ca, etc.) accounting for the remaining 8%. In petrographic calculations it is necessary to compare equal volumes of rock because the initial weathering (the saprolitization stage) causes little change in volume. The structure of most silicate rocks is such that 100 cations are associated with 160 oxygen ions. If some or all these cations are removed, the subsequent volume of material with 160 oxygen ions, contains fewer than 100 cations. The number of cations in some rock and mineral standard cells are given in Table 13.

Rocks near the surface of the earth have lost cations on weathering so that the cation total will be less than 100 (= more oxygen per cation = more highly oxidized). An exception is the mineral quartz which does not alter appreciably by weathering processes, although it is very slowly soluble.

The weathering of granite under tropical conditions in British Guiana (7°00′ N; 59°00′ W; warm, humid climate, 100-inch rain per year) was described by Harrison (1934), who made chemical analyses of the granite and its alteration products. Harrison's results are given in Table 14.

The ion ratios in a 160-oxygen-ion standard cell were calculated in from columns (c) and (d). The ratios in (d) have been adjusted so that Al (d) is the same number, 14.47, as in (c), that is, Al has been held static. If column (e) is substracted from column (d), the numbers in column (f) are obtained, indicating the loss and gain during weathering of the granite. Noteworthy in the weathering of this granite, as in the production of all weathering products, is the requirement for a large number of H^+ ions. As metal cations are moved out, H^+ ions replace them. A large amount of silica is removed. The standard cell of the fresh rock has 95.22 cations, and that of the weathered product has 79.11 cations. These figures are very similar to those of the average granite and average kaolinite cited earlier.

6–3. Calculations from Chemical Analyses

There has always been interest among soil scientists in finding out about the movement of the elements in rocks during soil formation. Many

TABLE 14

Granite Weathering to Kaolinitic Clay in British Guiana (after Harrison, 1934; Calculated by the Barth's Method, 1948)

	(a)	(b)		(c)	(d)	(e)	(f)
SiO_2	72.27	65.17	Si	63.07	52.61	37.60	−25.47
Al_2O_3	14.01	21.30	Al	14.47	20.25	14.47	
Fe_2O_3	1.95	1.38	Fe	1.78	1.21	0.86	− 0.92
FeO	0.60	0.53					
MgO	0.84	0.70	Mg	1.10	0.82	0.59	− 0.51
CaO	1.15	0.23	Ca	1.10	0.15	0.11	− 0.99
Na_2O	2.70	0.19	Na	4.51	0.29	0.21	− 4.30
K_2O	5.03	3.27	K	5.66	3.39	2.42	− 3.24
H_2O	0.52	7.13	H	3.04	39.37	27.42	+24.38
MnO	0.27		Mn	0.16			
TiO_2	0.58	0.65	Ti	0.37	0.39	0.29	− 0.08
P_2O_5	0.04		P				
			O	160.00	160.00	114.34	−45.66
	99.92	100.55					

(a) Analysis of fresh granite from quarry in British Guiana (Harrison, 1933; p. 62).
(b) Analysis of white sandy clay at top of quarry of (a) (Harrison, 1933; p. 62).
(c) Ions in 160-oxygen rock cell of (a).
(d) Ions in 160-oxygen rock cell of (b).
(e) Recalculated (d), holding Al constant to 14.47.
(f) Loss or gain of ions from (c) to (e).

years ago Hough, Gile, and Foster (1941), in describing soils developed on basalt in Hawaii, calculated the amounts of silica, alumina, and ferric iron lost or gained during profile development. Their results are given in Table 15.

Chemical analyses of soils developed on deeply weathered volcanic rocks (tuffs and lavas) on the island of Guam (Carroll and Hathaway, 1963) showed changes in the amounts of the major constituents under the leaching conditions that produced these soils. Figure 15 shows the approximate amounts of the major chemical constituents in bedrock and soil from one of the soil profiles examined in Guam.

It was first shown by Raggatt, Owen, and Hills (1945) that the removal of soluble constituents and the addition of water indicated chemically that the composition of a laterite or bauxite could be calculated from that of the parent material. This method was used by Carroll and Woof (1951) to describe a deep lateritic profile developed from basalt in northern New South Wales. From the results, a normative composition of the various horizons in the profile was calculated. The results are given in Table 16.

TABLE 15

Quantities of Constituents Gained or Lost by Different Layers of Soil in the Development of the Present Soil Profile (after Hough, Gile, and Foster, 1941)

Sample No.	*y* Depth, inches	*x* SiO_2, lb	*x* Al_2O_3, lb	*x* Fe_2O_3, lb	Original depth of layer, inches
C1643	0–10	+13	−243	−215	14.52
C1644	10–25	−63	+70	+177	13.08
C1645	25–40	+50	+173	+37	12.40
C1649	0–10	−172	−392	−680	23.05
C1650	10–25	+34	+138	+431	8.79
C1651	25–40	+183	+253	+243	8.16
C942	0–7	−43	−130	−140	10.21
C943	7–19	−11	−12	−31	12.52
C944	19–31	+53	+142	+171	8.27
B5787	0–6	−107	−256	−205	11.86
B5788	6–9	−36	−86	−50	4.78
B5789	9–30	+19	+85	+231	17.72
B5790	30–40	+124	+262	+24	5.63
C938	0–7	−270	−351	−190	15.18
C939	7–15	−161	−124	+117	9.65
C940	15–22	+199	+222	+36	2.40
C941	22–30	+241	+253	+38	2.75

x) Pounds gained or lost per unit area such that 1 inch = 100 lb; the assumed thicknesses of the unit areas are given in the last column.
y) Present depths of layers; it is assumed that layers showing a net loss of constituents were originally thicker, and layers showing a net gain of constituents were originally thinner.

A red-yellow podzolic soil named monalto, developed on metagabbro near Bel Air, north of Baltimore, Maryland, was examined chemically and mineralogically. The chemical analyses of the soil and rock were used to calculate the results of weathering by the method of Raggatt, Owen, and Hills. The results of the calculations are given in Table 17.

6–4. Calculations from Mineralogical Analyses

In another kind of petrographic calculation, Brewer (1955) made a detailed mineralogic examination of a podzolic soil that had developed a

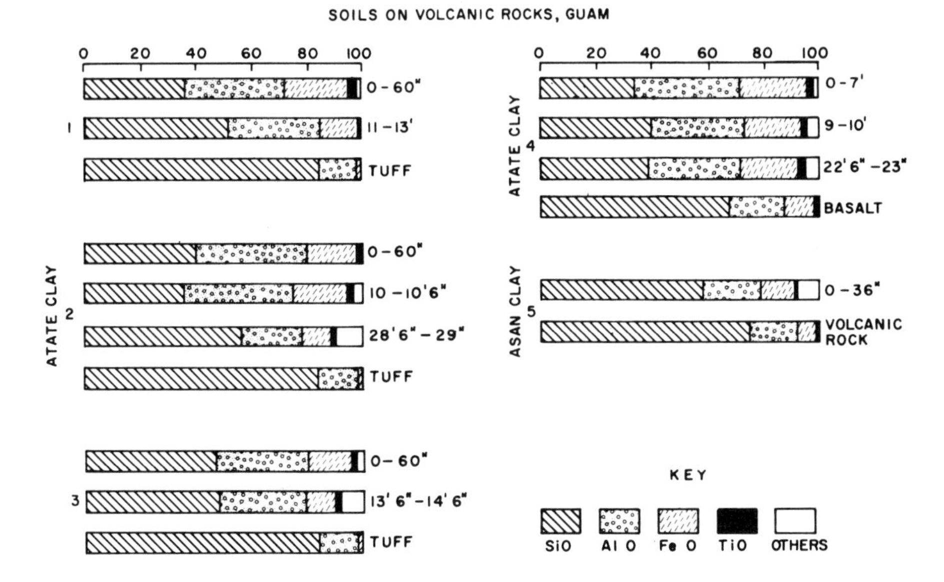

Fig. 15. Chemical composition of soil and underlying rock in Guam (after Carroll and Hathaway, 1963).

TABLE 16

Normative Mineralogic Composition (Weighted Average) of Soil Passing a 2-mm Sieve from a Lateritic Profile on Basalt in Northern New South Wales (Carroll and Woof, 1951; p. 92)

Sample depth, ft	Kaolinite, %	Gibbsite, %	Hematite, %	Ilmenite, %	Anatase, %	Leucoxene, %	Remainder, %
0–4	19	58	8	8	4	—	3
4–9	21	61	4	6	—	8	—
9–10.5	22	50	3	4	—	15	6*
10.5 –11.5	22	37	2	2	—	36	1
11.5 –12.25	6	90	—	—	3	—	1
12.25–16.25	72	15	1	—	6	—	16
16.25–23+	52	23	6	2	2	—	15†

* Includes 1% goethite.
† Excess alumina should be allocated with Fe to nontronite, gibbsite thus being reduced in amount.

TABLE 17

Comparison on a Water-Free Basis of the Chemical Composition of Metagabbro and of the Soil Profile Developed above It from near Bel Air, Maryland (after Carroll, 1953; p. 98)

<table>
<tr><th></th><th>(a)</th><th>(b)</th><th>(c)</th><th>(d)</th><th>(e)</th><th>(f)</th><th>(g)</th><th>(h)</th><th>(i)</th><th>(j)</th></tr>
<tr><td>SiO_2</td><td>51.5</td><td>0</td><td>51.5</td><td>62.0</td><td>47.9</td><td>−14.2</td><td>54.0</td><td>−8.0</td><td>45.4</td><td>−16.6</td></tr>
<tr><td>Al_2O_3</td><td>17.3</td><td>0</td><td>17.3</td><td>20.9</td><td>27.4</td><td>+ 6.5</td><td>22.7</td><td>+1.8</td><td>28.6</td><td>+ 7.7</td></tr>
<tr><td>Fe_2O_3</td><td>2.7</td><td rowspan="2">0</td><td rowspan="2">12.8</td><td rowspan="2">15.5</td><td rowspan="2">14.8</td><td rowspan="2">− 1.2</td><td rowspan="2">15.5</td><td rowspan="2">nil</td><td rowspan="2">14.5</td><td rowspan="2">− 1.0</td></tr>
<tr><td>FeO</td><td>8.6</td></tr>
<tr><td>MgO</td><td>4.7</td><td>4.7</td><td>0</td><td>—</td><td>—</td><td>—</td><td>—</td><td>—</td><td>—</td><td>—</td></tr>
<tr><td>CaO</td><td>13.8</td><td>13.8</td><td>0</td><td>—</td><td>—</td><td>—</td><td>—</td><td>—</td><td>—</td><td>—</td></tr>
<tr><td>TiO_2</td><td>1.2</td><td>0</td><td>1.2</td><td>1.5</td><td>0.1</td><td>− 1.4</td><td>0.2</td><td>− 1.3</td><td>—</td><td>—</td></tr>
</table>

The figures shown are percentages of the weighted average.
(a) Metagabbro recalculated to 100%.
(b) Presumed losses from (a).
(c) Remainder in profile.
(d) Remainder calculated to 100%.
(e) Whole profile, weighted average.
(f) Difference, (d) minus (e).
(g) Profile 0–32 inches, weighted average.
(h) Difference, (g) minus (d).
(i) Profile 32–98 inches, weighted average.
(j) Difference, (i) minus (d).

profile 174 inches deep above a saprolitic granodiorite near Braidwood, New South Wales (35°30′ S; 149°47′ E). The parent rock was so deeply weathered that no fresh rock was obtainable. It was found that the following mineralogical changes occurred:

Feldspar (orthoclase and plagioclases)	Kaolinite with 40% weight loss
Hornblende	Kaolinite with 25% weight loss
Biotite	Kaolinite with 70% weight loss
Quartz	Unaltered

From these reactions the weight of kaolinite, and therefore the percentage of kaolinite, that can be formed by weathering *in situ* under these conditions (upland, 26.5 inches of annual rainfall evenly distributed) can be calculated. If complex clay minerals are formed there may be no loss of weight in the formation of clay minerals from primary minerals. In this profile montmorillonite is the first clay mineral to be formed when the rock weathers; montmorillonite progressively changes to kaolinite towards the upper part of the profile. The results of this investigation are schematically shown in Fig. 16.

In the mineralogical method proposed by C. E. Marshall, a quantitative determination is made in every part of the soil profile of a mineral that is resistant to weathering and is at the same time immobile. Such a mineral is found in the nonclay fraction of the soil. Marshall termed this mineral an *index mineral.* As we have seen in Section 4–3, the common minerals can be arranged in a stability series, and in any such series the most stable minerals are zircon and tourmaline. The presence of certain kinds of rocks in an area may suggest that other minerals be used. For example, in examining a number of old residual sands on a Precambrian shield, Carroll (1939) found that the metamorphic minerals, staurolite, kyanite, sillimanite, and andalusite had to be used as index minerals. In other instances, the amount of heavy residue from the sand fraction was used. Barshad suggests quartz, albite, and microcline. Haseman and Marshall (1945) used zircon for a Grundy silt-loam profile, as did Whiteside and Marshall (1944) for a Putnam silt-loam profile.

6–5. Calculating the Amount of Clay Developed

Probably the most important feature of weathering is the formation of clay minerals. Part of the clay may originate in the deeply weathered rock, but most of it forms during pedochemical weathering. There are two principal methods that have been used for determining the amount of clay that has formed from a rock during weathering. One method is mineralogical, the other chemical. Barshad (1964) discussed a way of obtaining results from both methods.

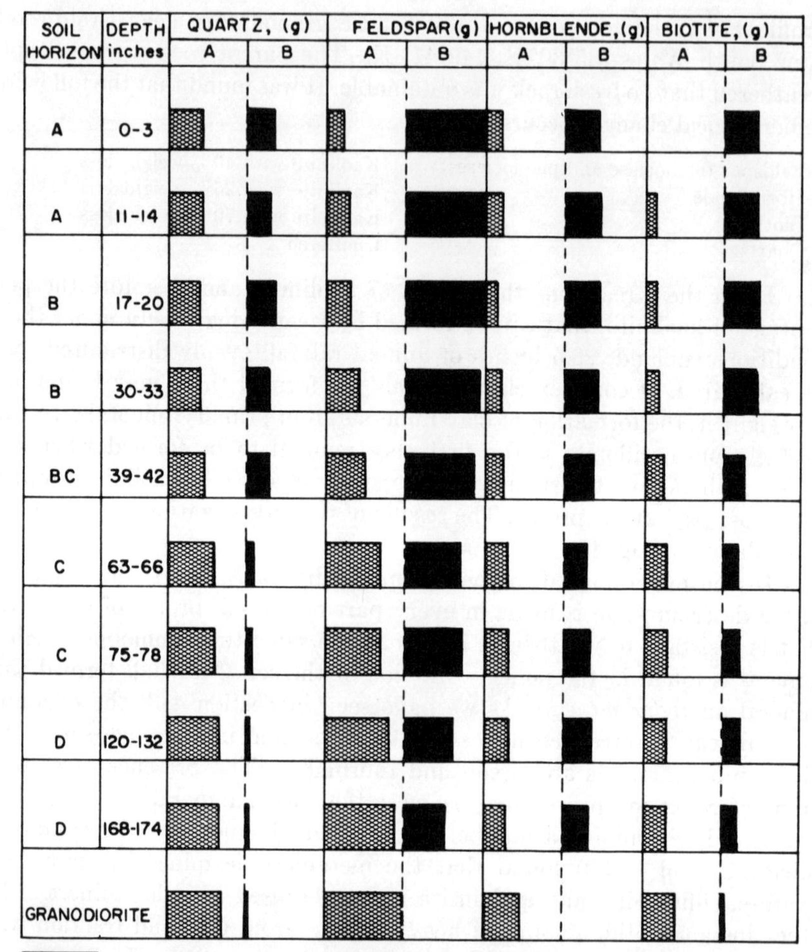

Fig. 16. Weight of each mineral in soil derived from granodiorite (after Brewer, 1955).

In using the index-mineral method of calculating the amount of clay developed in a soil profile, one has to assume a uniform distribution of the stable mineral in the parent material, and in order to do this the investigator should determine the distribution of such a mineral in the parent rock. Another feature that should be taken into consideration is the varietal character of the mineral. For example, a soil could contain about the

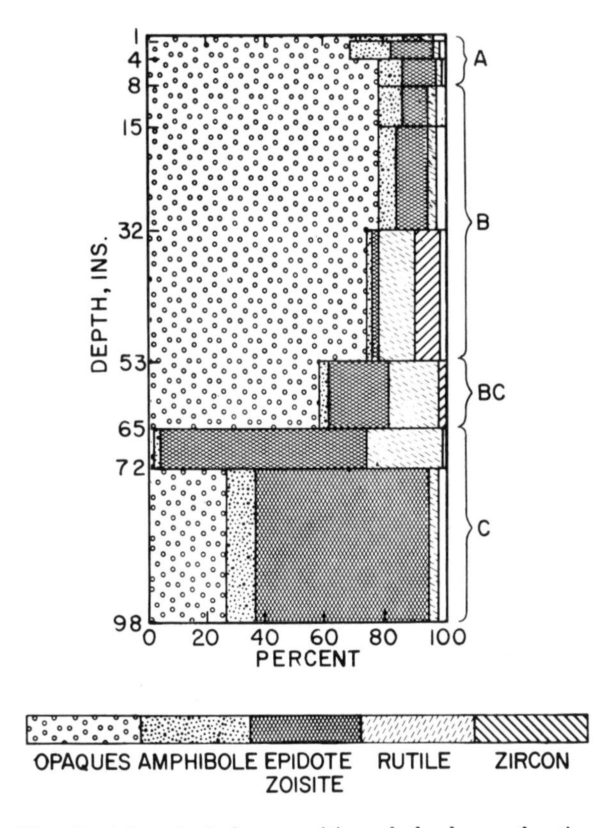

Fig. 17. Mineralogical composition of the heavy fraction (sp. gr. <2.9) of the very-fine-sand fraction of a monalto soil at Bel Air, Maryland (after Carroll, 1953).

amount of zircon expected from a certain rock, but the zircon itself may be from different sources and therefore look different; the parent rock may contain all clear, colorless, sharply angular zircons, whereas the soil contains some clear, colorless zircons, and some rounded, purple ones. In a chemical analysis for zirconium, such differences would not be known, but a mineralogist would know that the soil material had not all been derived from the suggested parent rock. Another feature that could be brought out by a mineralogical examination would be the variation in amount and kind of the heavy minerals. This would suggest that the soil profile, although it appeared to have developed from a single rock, was in reality a composite profile. An example of this is the monalto profile examined by Carroll (1953), in which the uppermost portion was derived from a thin veneer of a much later formation. Figure 17 illustrates this.

TABLE 18

Quantitative Evaluation, Using Chemical Analyses, of Clay Formation, Clay Migration, and Volume Changes during Development of Soil Material of the A_2 and B Horizons of a Dayton Silt-Loam Profile (after Barshad, 1964; p. 27)

Horizon	Present volume per horizon, cc	Bulk density, g/cc	Weight of present soil per horizon, g	Weight of parent material per horizon, g	Original volume per horizon, cc	Change in volume, cc	Relative clay loss or gain, g*	Relative clay loss or gain, g†	Relative change in volume, cc‡
A_2	10.2	1.22	13.4	19.4	14.0	−3.8	−35.7	−25.7	−27.1
B	10.2	1.56	15.9	12.3	8.8	+1.4	+45.2	+32.3	+15.9
C	20.3	1.39	28.2	28.2	20.3	0.0	0.0	0.0	0.0

* Per 100 cc of parent material.
† Per 100 g of parent material.
‡ Per 100 cc of parent material.

Using this method, Barshad (1964, p. 16) found that the following amounts of clay were formed in a Sheridan clay-loam profile, a brown soil in the vicinity of Davis, California:

Horizon	*Clay formed per cc of parent material, g*
A_1	7.2
A_2	7.5
B_1	5.4
B_2	0.8
B_3	—
C	—

In the chemical method of determining the amount of clay formed, it is assumed that the clay fraction ($<2\ \mu$) is the product of weathering, whereas the nonclay fractions are the reactants. Therefore, the amount of clay formed must be proportional to the loss of those minerals of the nonclay fraction which alter to clay. In young soil profiles (immature) developed from a uniform parent material, the clay content is constant with depth. In moderately weathered soils on granitic or basaltic rocks, the clay content is highest at the surface and gradually decreases with depth. Since chemical composition of a material reflects its mineralogical composition, the amount of change in the chemical composition of the nonclay fraction of the parent material, as a result of soil development, should also reflect the amount of clay formed. Such a chemical analysis can be utilized to calculate the amount of clay formed. This is a useful method, as chemical analyses have been made for many soils in various parts of the world and these analyses could be used for calculating soil-profile development under various conditions. Barshad (1964) gives details of the steps involved in these calculations. The steps are as follows:

1. Calculating the chemical composition of the nonclay fraction.
2. Calculating the amount of clay formed in any horizon from 100 g of nonclay of the parent material.
3. Calculating the amount of clay formed in the A horizon which resulted in the present amount of nonclay in 100 g of soil.
4. Calculating the amount of clay originally present in the A horizon.
5. Calculating the loss or gain in clay at each horizon during soil-profile development.
6. Calculating the change in volume of each horizon during soil development.
7. Calculating the change in the total chemical composition of the soil at each horizon during soil development.

Barshad (1964, p. 27) gives the results of such a series of calculations of clay formation in a Dayton silt-loam profile in Table 18.

Chapter 7

PHYSICAL WEATHERING

Physical weathering takes place when solid rocks are broken into fragments with little or no chemical change in the rock itself. Only physical weathering occurs in very cold, dry or very hot, dry climates, and splitting is largely due to insolation. A chemical analysis of the fragmented rock would be similar to that of the solid rock. Disintegration of rocks in moist climates accompanies chemical weathering.

7–1. Mechanical Disintegration

Mechanical disintegration of rocks without any chemical changes being produced in the mineral components only takes place in very arid regions, or where freezing conditions cause splitting and fracturing. The first stage in disintegration is caused by the jointing present in most rocks. Such jointing is largely due to the removal of pressure when rocks become exposed at the earth's surface.

The effects of disintegration differ in the various kinds of rocks, but the net result is the production of a loose granular mass of mineral material from a compact rock.

7–2. Primary Breakdown into Grains

In igneous and metamorphic rocks the grain size, and the way the grains are packed together in the rock, influences how the individual minerals will break into different sizes. A coarsely granular rock will disintegrate more rapidly than one of finer grain; this is due to the thermal expansion of the individual minerals. Sedimentary rocks, however, generally contain grains within narrow ranges of size. In sandstones and quartzites, for example, grains tend to be monomineralogic, but the cement commonly present in many sedimentary rocks binding the grains together influences

TABLE 19

Monomineralogic Grains in Disintegrated Granitic and Gneissic Rocks (after Pettijohn, 1957)

Grain size, mm	Granite, number of grains	Percent	Gneissic quartz diorite, number of grains	Percent
8–4	—	—	—	—
4–2	17	6.27	—	—
2–1	36	12.36	10	4.54
1–0.5	64	21.76	42	19.09
0.5–0.25	80	27.49	78	35.45
0.25–0.12	94	32.30	90	40.91

the size and shape of the disintegration product. Composite grains are formed from the finer-grained sedimentary and igneous rocks, but coarse-grained rocks yield grains of individual minerals on disintegration. The percentage of monomineralogic grains in disintegrated granite and gneiss found by Pettijohn (1957) is given in Table 19.

The size distribution of grains of disintegrated igneous rocks has been shown by Krumbein and Tisdel (1940) to be rather similar to that of artificially crushed stone. The grain-size distribution of disintegrated sedimentary rocks is similar to that of the original sediment, that is, the size of grains in a disintegrated sandstone is similar to that of the original sand of which the rock was made. Thus, the parent rock is important physically as a source of the skeletal or framework material of a soil. Those rocks that contain only a small quantity of the larger sizes of material cannot produce coarse-grained soils. For example, a pure limestone produces a clayey soil as the calcium carbonate is gradually removed.

The amount of coarse material formed from the weathering of sedimentary rocks can be estimated from the acid-insoluble residues. Table 20 gives the amounts of insoluble residue of some sedimentary rocks of the Paleozoic sequence in southwestern Virginia (Carroll, 1959a).

In Table 20 some of the rocks are derived from sediments (sandstones and shales) and some from an admixture of detrital material in a certain depositional environment (sandy limestones); others are chemically deposited (limestones) or diagenetically altered (dolomites, cherty limestones).

It is more difficult to assess the amount of insoluble material due to mechanical disintegration of igneous and metamorphic rocks than it is

TABLE 20

Insoluble Residues of Some Sedimentary Rocks (Carroll, 1959[a])

Rock	Percent insoluble residue	Pounds per acre-inch
Sandstone (Chemung)	90	531,400
Shale (Brailler)	87	511,750
Shale (Martinsburg)	30	176,470
Shale (Martinsburg)	40	235,300
Limestone (Lenoir)	0.8	4,900
Limestone (Mosheim)	0.5	3,000
Limestone (Beekmantown, dolomitic)	0.2	1,000
Limestone (Beekmantown, cherty)	13.5	82,500
Dolomite (Beekmantown)	1.7	10,400
Limestone (Conococheague)	40	235,400
Limestone (Elbrook)	1.6	10,130

for sedimentary rocks. However, from a knowledge of the mineralogical make-up of igneous rocks (Section 1–1) one can predict the amount of insoluble material that will remain after the first stage of chemical weathering. In general, the more highly siliceous the rock, the greater the amount of quartz present. On weathering, a granite will produce a greater amount of insoluble material than a basalt. The amount of residue can be arranged in the following order from the rocks listed in Section 1–1:

$$\text{granite} > \text{syenite} > \text{diorite} > \text{gabbro} > \text{basalt} > \text{peridotite}$$

The grain sizes found in the primary breakdown of rocks are given in Table 21. Included in this table are the kinds of rocks for which data are available and the climate (according to Köppen) under which each rock weathered.

The sizes of the grains (measured with standard sieves) in Table 21 have been grouped into coarse sand, sand, silt, and clay. The classification used follows the Wentworth scale (1922) which is a scale of grade terms very generally used for describing the grain sizes of sedimentary materials. Three commonly used classifications of grade sizes are given in Table 22.

In describing the first stages of disintegration of rocks, it is difficult to separate physical disintegration from the effects of chemical alteration. For example, when the iron present changes from ferrous to ferric on oxidation, the alteration produces cracks and weaknesses which cause disintegration to proceed in certain directions, particularly along grain boundaries. Recent investigations in Antarctica (Kelly and Zumberge,

TABLE 21
Grain Size of Disintegrating Rocks at the First Stage of Weathering

Rock	Locality	Climate (Fig. 3, Section 2–2)	Coarse sand* (>1.0 mm), %	Sand† (1.0–0.1 mm), %	Silt (0.1–0.002 mm), %	Clay (0.002 mm), %	Reference
Gneiss	Ghana	Am	21.3	33.4	9.2	30.0	Stephen, 1953.
Andesite	Grenada	Am	—	87.4	9.7	2.9	Hardy and Rodrigues, 1939a.
Volcanic agglomerate	Guam	Am	2.4	10.7	19.2	67.7	Carroll and Hathaway, 1963.
Basalt	Kauai	Am	4.8	23.5	30.5	41.2	Hough and Byers, 1937.
Serpentine	Cuba	Af	2.2	11.1	43.0	43.7	Bennett and Allison, 1928.
Andesite tuff	Puerto Rico	Amw	—	8.0	13.0	70.0	Bonnet, 1939.
Gneiss	Ceylon	Aw	75.8	19.0	1.9	3.3	Pannebokke, 1959.
Basalt	Maui	Aw	0.6	25.6	37.3	36.7	Hough and Byers, 1937.
Dolerite	Natal	Aw	8.9	17.7	16.7	56.6	Beater, 1947.
Basalt	Norfolk Island	Aw	—	17.0	26.0	55.0	Stephens and Hutton, 1954.
Basalt	Colorado	Bw	28.8	40.7	15.8	15.0	Short, 1961.
Basalt	Queensland	Cw	—	13.7	17.5	57.1	Tamhane and Namjoshi, 1959.
Gneiss	W. Australia	Csa	1.1	39.0	6.6	53.1	Carroll, unpublished.
Basalt	S. Australia	Csb	13.0	32.0	48.0	8.0	Tiller, 1958.
Granodiorite	Japan	Cfa	24.9	68.3	5.3	1.5	Yamasaki et al., 1955.
Meta-andesite	N. Carolina	Cfa	—	18.5	66.5	15.0	Short, 1961.
Diabase	N. Carolina	Cfa	48.8	47.2	0.6	3.4	Hardy and Rodrigues, 1939b.
Granite	Germany	Cfb	67.0	21.0	10.0	2.0	Preusse, 1957.
Basalt	Germany	Cfb	9.2	72.4	5.6	12.8	Huffman, 1954.
Granodiorite	New South Wales	Cfb	29.2	65.8	4.1	0.9	Brewer, 1955.
Muscovite schist	Virginia	Cfb	15.3	33.6	4.2	46.6	Rich and Obenshain, 1955.
Granite	Wyoming	Dfb	41.1	53.5	3.4	1.9	Short, 1961.
Syenite	Norway	Dfc	53.5	11.5	3.4	1.0	Holtedahl, 1953.

The figures were abstracted from the literature by the author.
Grade terms follow the Wentworth scale (1922).
* Includes all granular material coarser than 1-mm grain diameter.
† Includes medium, fine, and very fine sand.

TABLE 22

Three Commonly Used Classifications of Grade Sizes of Soils

Wentworth scale	mm*	U.S. Department of Agriculture	mm*	International grade classes	mm*
Gravel	4–2	Stones	>2.0		
Very coarse sand	1–2	Gravel	1–2		
Coarse sand	0.5–1		0.5–1	Coarse sand	2–0.2
Medium sand	0.25–0.5		0.25–0.5		
Fine sand	0.12–0.25		0.25–0.10	Fine sand	0.2–0.02
Very fine sand	0.12–0.062		0.10–0.05		
Silt	0.062–0.002		0.05–0.002	Silt	0.02–0.002
Clay	<0.002		<0.002	Clay	<0.002

* Grain diameter by standard sieves; clay and silt by sedimentation, using Stokes' law.

1961) show that physical disintegration is the most important process of weathering in this region, an opinion that has always been accepted by geologists. The weathering of a fine-grained diorite they studied results in the production of fine sand of the same chemical composition as the fresh rock. The only chemical change was oxidation of the ferrous iron to a limonite which colored the sand brown, whereas the rock was light gray in color.

The results of weathering in the Antarctic climate were further studied by Tedrow and Ugolini (1966) who found that the older glacially derived materials, such as moraines, developed, under the abiotic conditions prevailing, a kind of soil profile in which the surface layer (horizon) of the older material has a brownish color, whereas the younger material is uniformly gray. Figure 18 illustrates a typical profile of this material.

The grain-size analyses of two soils from the base of Cape Hallett, Antarctica were reported by Rudolph (1966) and are given in Table 23.

7–3. Effect of Particle Size on Weathering Processes

Inasmuch as chemical weathering is mainly the result of surface reactions between solutions and mineral grains, any increase in the surface area available for chemical reactions is very important. Baver (1956, p. 12) has calculated that if a cube with faces 1 cm^2 (100 mm^2) is broken down into smaller and smaller cubes, the area of fine sand-sized particles (about 0.25–0.125 mm diameter) will be 314 cm^2 (about 3140 mm^2), the

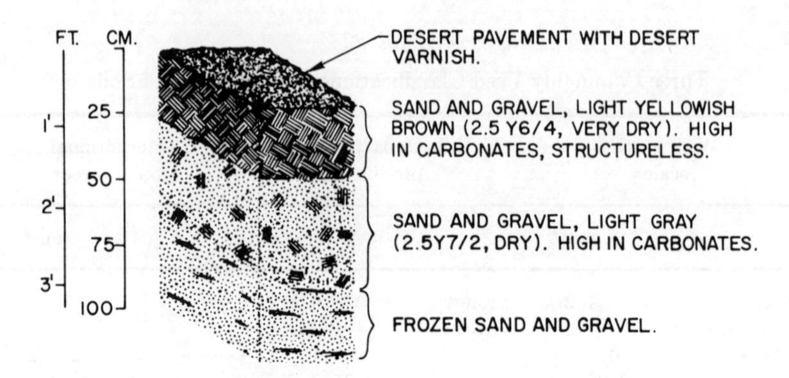

Fig. 18. Idealized abiotic soil of the cold desert (after Tedrow and Ugolini, 1966).

area of silt-sized particles (about 0.125–0.062 mm diameter) will be 1570 cm^2 (about 15,700 mm^2), the area of claysized particles ($<62\ \mu$) will be 31, 416 cm^2 (314,160 mm^2), and the area of colloidal clay ($<50\ \mu$) will be 628,320 cm^2 (6,283,200 mm^2). Thus, the clay fraction of a weathering product has 100 times the surface area of the fine sand on which reactions can take place. This is one of the fundamentals in weathering.

In igneous and metamorphic rocks the grain size and packing influences the disintegration, as the minerals will break into different sizes. A coarsely granular rock will disintegrate more rapidly than one of finer grain; this is partly due to the thermal expansion of the individual minerals. Sedimentary rocks, however, generally contain grains within narrow ranges of size and they tend to be monomineralogic, although cement binding the grains influences the final product. Composite grains will be formed from

TABLE 23

Grain-Size Analyses of Two Soils at Cape Hallett, Antarctica (after Rudolph, 1966)

Grain size, mm	Soil in experimental grass-plot area, %	Soil in lichen area, %
Very coarse sand (2–1)	15.7	55.1
Coarse sand (1–0.5)	55.2	33.7
Medium sand (0.5–0.25)	19.9	4.6
Fine sand (0.25–0.1)	8.4	1.7
Very fine sand (0.1–0.05)	0.3	0.4
Silt (0.05–0.002)	0.2	3.0
Clay (<0.002)	0.3	1.5
Fine clay (<0.0002)	tr.	0.3

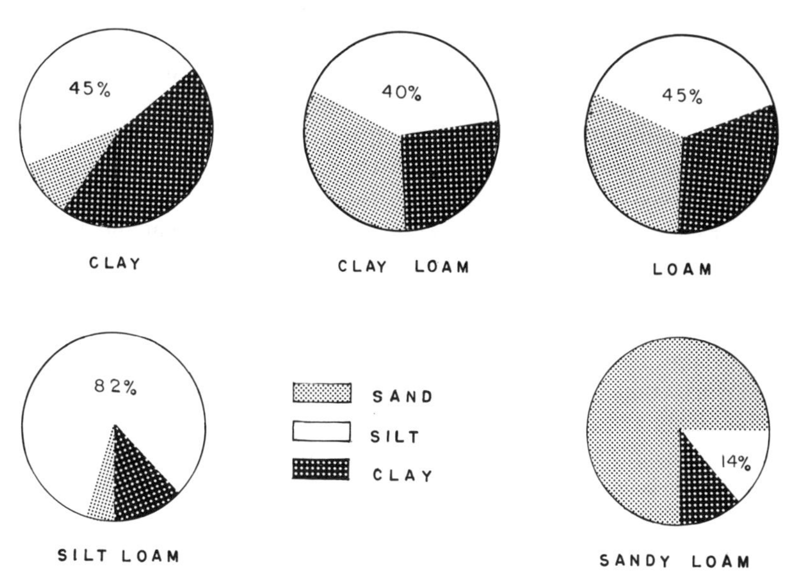

Fig. 19. Amounts of sand, silt, and clay in common soil types (after Rice and Alexander, 1938).

the finer grained sedimentary and igneous rocks, but the coarser grained igneous and metamorphic rocks yield grains of single minerals.

7–4. Grain Sizes in Soils

The grain size of the ultimate products of weathering, soils, is important for several reasons. As noted in Section 7–3, as mineral grains are reduced in size by splitting along cracks and mineral cleavage planes the surface area is enlarged, thereby exposing a larger amount of the mineral to chemical processes. The grains of disintegrated rock form a framework or skeleton which makes up and supports the body of the inorganic weathering product. This framework consists of small rock fragments and grains of individual minerals. If there has been considerable chemical alteration of the easily altered minerals of a parent rock then some clay will have formed as an alteration product. Clay is colloidal in size, and when present it fills up the spaces between the partially or completely larger mineral grains and unweathered rock fragments, as shown in Plate 2, A and B. The relationship between the framework and the fine material, the clay, is important; if the pore spaces in the weathered material are filled up, water and air cannot enter the pores. Both are required for the growth of plants which is essential to the development of most soils. The composition of water in the weathering material is dependent on the rainfall of the region and also on the carbon dioxide absorbed from the soil air, the composition of which is related to the metabolic processes of macroorganisms and microorganisms in the soil. Pore space, influenced by the sizes of the grains, allows water to drain through the soil if the climate is sufficiently moist. This water carries with it chemical elements dissolved from the partial or complete solution of minerals. In other words, chemical winnowing occurs which results, in geologic time, in leaving insoluble minerals in the soil framework and carrying away in groundwater, streams, and rivers all the soluble chemical elements. Analyses of water from rivers draining large land areas indicate how much chemical denudation of the land surfaces takes place by this means. The sizes of the particles in soils and the amount of each size is a measure of the kind of rock that was the parent of the soil, and of the weathering that has taken place. The distribution of particle sizes in saprolitic rock is different from that in soils; further description of the particle sizes in soils is given in Chapter 4 and Fig. 19. Descriptive terms that are in common usage for the principal types of soil are given in Fig. 19 which also shows the average amounts of sand (2–0.2 mm), silt (0.2–0.002 mm) and clay (<0.002 mm) to be expected in these soils.

Chapter 8

CHEMICAL WEATHERING

The earth's surface is soil developed from weathered rock except in the highest mountains, glaciated areas, and deserts. Each kind of rock is weathered according to its composition, texture, and to the chemical environment which is within the framework of known pH and Eh conditions (Section 2–1) of the region in which it occurs. Regional patterns of weathering can be mapped as in Fig. 3 (Section 2–2). Weathering is a series of surface chemical reactions between rocks, the atmosphere, and water; it is a chemical process of silicate reactions to leaching by water at low temperatures (30°C) and in atmospheric pressure ranging from that at sea level to that of the highest mountains. The end products of weathering are oxidized residuum from rocks and chemical solutions containing the soluble chemical elements of the original rocks. Chemical weathering can be represented thus:

$$\text{Rock} + \text{oxygen} + \text{water} + p\int_{\text{mountains}}^{\text{sea level}} \rightarrow \text{saprolite} \rightarrow \text{soil} + \begin{array}{l}\text{soluble}\\ \text{inorganic}\\ \text{and organic}\\ \text{ions or com-}\\ \text{pounds}\end{array}$$

Although the principal processes of chemical weathering appear simple, these processes are affected by numerous variable factors, so that a diversified pattern of weathering products is formed (see Fig. 10, Section 5–3).

8–1. Weathering by Water

Rain supplies the water that causes the chemical weathering of rocks. Water is the solvent of the silica in the framework of the aluminosilicate minerals which form the mass of the rocks at the earth's surface, and without water there could only be physical disintegration and a limited amount of oxidation. Geochemical weathering (Section 3–1) is inorganic, whereas pedochemical weathering (Section 3–2) is the result of both organisms and organic solutions reacting with the results of inorganic-solution reactions with rocks. It has been well said that "*Das Wasser mit dem gelösten Nährstoffen ist für Boden und Untergrund das, was das Blut für den lebenden Organismus ist: ohne Wasser, kein Boden. Deshalb ist seine Wasserführung von grösster Bedeutung für die Bodenbildung.*"

In a very general way weathering can be conceived as an acid attack originating at the lithosphere–atmosphere contact (personal communication, Garrels, 1957). Rainwater containing CO_2 is progressively neutralized as it percolates downward. The rock minerals take up hydrogen ions and water, and release cations to the transient water. When pedochemical weathering follows geochemical weathering (Section 3–2) in the soil zone, additional CO_2, as well as gases and organic acids, are produced both by the decomposition of organic material and by the metabolic processes of living organisms, that is, by the microorganisms and by the vegetative cover. The products of mineral alteration migrate downward in solution, as colloids, and by differential physical movement of coarse and fine resistant particles. There is a tendency toward the formation of layers (the soil horizons) containing relatively few minerals, and with the resistates and ultimate hydrolyzates concentrated in the upper layers. In silicate rocks, sheet silicates containing small percentages of cations form a zone of intermediate alteration above primary three-dimensional silicates.

Weathering depends on the chemical characteristics of the elements combined to form minerals in rocks. In Chapter 7 we saw how rocks break up into smaller particles, thus exposing larger areas of their constituent minerals to alternation by reactions with their environment. Because each chemical element has an ionic radius and ionic charge it reacts differently to weathering agents. Figure 20 groups the common cations according to their ionic radii and charge, which determine their reaction on weathering.

These elements are bonded in various ways to form minerals. Weathering begins when the weakest bond in a mineral is made to change or disappear. There are four types of bonding in the common minerals:

1. In structures with independent silica tetrahedrons or with linked tetrahedra, the weakest bond binds the tetrahedrons together; quartz is an example.
2. In metasilicates containing single or double chains of linked tetrahedra (pyroxenes and amphiboles), the weakest bond binds the chains together.
3. In structures with two-dimensional tetrahedral linkage (the sheet

structure in micas), the weakest bond binds the bases of the tetrahedrons together; the K ion in mica is an example.

4. In structures with three-dimensional tetrahedral linkage, the weakest bond binds the cations that balance the charge of the alumina tetrahedrons; the K^+, Na^+, or Ca^{+2} in feldspars.

The different types of bonding have enabled the observed weathering sequences (Chapter 4) to be formulated. Table 24 gives the common reactions that occur during weathering.

Minerals yield a considerable amount of chemical compounds on weathering. A biotite separated from a granite was analyzed; the amounts of oxides that 1000 lb of this biotite contains and the distribution of these oxides on weathering are given in Table 25.

As rocks and minerals are altered by chemical weathering the elements of which they are composed are accommodated in the weathering products or taken away in drainage waters. The general chemical effect in a moist climate (leaching conditions, pedalfer soils) is removal of soluble cations from minerals thereby producing a simplified mineral structure, one that is in equilibrium with the temperature and pressure at the earth's surface. For example, in the change of potassium feldspar to kaolinite the feldspar structure is destroyed, its place being taken by that of kaolinite with a 1:1 layer structure. As potassium is removed fine mica known as sericite

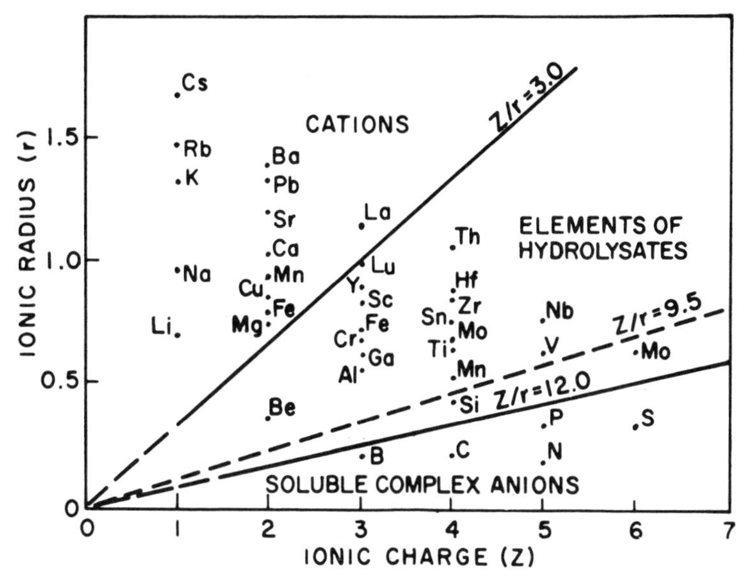

Fig. 20. Grouping of certain elements according to their ionic radii and charges (after Gordon, Tracey, and Ellis, 1958).

TABLE 24

Common Chemical Reactions that Take Place during Weathering

Element in mineral	Process	Product	Presence of air	Presence of water	Product
Fe^{+2}	Oxidation	Fe^{+3}	+	–	Hematite (mineral)
Fe^{+2}	Reduction	Fe^{+2}	–	+	$Fe(OH)_2$, in solution
Ca^{+2}	Acidification in carbonates*	Ca^{+2}	–	+	$Ca(OH)_2$, in solution
Ca^{+2}	Acidification in silicates*	Ca^{+2}	–	+	Ca^{+2}, in solution
Mg^{+2}	Acidification in carbonates*	Mg^{+2}	–	+	$Mg^{+2}x$, in solution
Mg^{+2}	Acidification in silicates*	Mg^{+2}	–	+	Mg^{+2}, in solution
Na^{+}	Solution*	Na^{+}	–	+	Na^{+}, in solution
K^{+}	Ion exchange	K^{+}	–	+	K^{+}, in solution
Al^{+3}	Hydrolyzation	Al^{+3}	–	+	$Al(OH)_3$ mineral; alumina is insoluble between pH 4 and 8.
Si^{+4}	Solution†	Si^{+4}	–	+	Si^{+4}, slightly soluble

All these elements with the possible exception of potassium and silicon form chelates with organic acids.

* Calcium and magnesium ions also enter into ion-exchange reactions.

† Silicon in quartz is soluble to the extent of 6 ppm at all pH values; silicon combined in other minerals is soluble to the extent of 140 ppm at all pH values.

TABLE 25

Amounts of Chemical Constituents in 1000 lb of Biotite Separated from a Granite

Chemical compound	Amount per 1000 lb of biotite, lb	Distribution of weathering
SiO_2	410	In unweathered mica flakes, clay minerals, and leached by drainage water
Al_2O_3	167	In unweathered mica flakes, but largely in clay minerals
Fe_2O_3	38	Most as iron-oxide coating on other grains
FeO	165	Oxidizes to Fe_2O_3 as mica is weathered and is dispersed to give brown color to clay minerals and coatings of very fine iron oxide on other grains
MnO	4	In clay minerals in isomorphous replacements; finely disseminated with iron oxides
MgO	54	Largely removed in drainage water, but some remains in unaltered mica and in clay minerals
CaO	27	As for MgO, but is not leached in arid climates
Na_2O	11	Removed in drainage water
K_2O	71	Remains part of mica until mica is converted to clay minerals by weathering; gradually leached out
TiO_2	17	Originally part of the mica structure; recrystallizes as anatase, brookite, or sphene on weathering; becomes a stable mineral in the weathering product
P_2O_5	<1	Absorbed by clay minerals, particularly alumina
F	4	Leached out, but some remains with unaltered mica
H_2O + and −	31	In clay minerals; some may remain in unweathered mica
FeS_2	<1	Oxidizes and produces iron oxides and SO_2
Cr_2O_3	<1	On weathering of mica remains as discrete rains of chromite
V_2O_5	<1	Oxidizes to tetravalent form and remains with the clay minerals

The separations and calculations are by the author; the analysis of the biotite is by the Mines Department, Western Australia.

forms, this sometimes being the first indication of alteration of feldspar. In a dry climate the soluble chemical elements cannot be completely removed from the original minerals. Weathering is very slow, and those elements that are moved from their original positions in a mineral crystal lattice will accumulate in the soil formed (nonleaching conditions, pedocal soils). The climatic environment steady state determines the kind of weathering that occurs in any region.

8–2. Kinds of Water in Weathering

The principal source of water is rain, but in weathering the water present in rocks and soils, groundwater and soil water, is the true reacting electrolyte. Rainwater (and snow at high altitudes and latitudes) is the primary source of cations and anions that are added to the surface of the lithosphere, except in desert areas where dry precipitation is important. The water, modified by its situation as groundwater or soil water, reacts with rocks and minerals, and thereby releases chemical elements that either continue the weathering process or are removed by drainage. At one time rainwater used to be considered pure water, rather like distilled water; but now that numerous chemical analyses of rainwater have been made, it is known to have a variable chemical composition. This composition is seasonally stable in most environments (Carroll, 1962). Table 26 gives the chemical composition of rainwater in different countries.

The pH of rainwater has a seasonal and regional pattern. Rainwater in equilibrium with the CO_2 of the atmosphere (as it is in clouds) has a pH of 5.7, which is slightly acid. In northern Europe the average pH of rainwater is 5.47; in the United States, the pH values are between 6 and 7. In tropical regions the pH of rainwater is a little above 7, but in areas of active volcanoes it may be below 4, as has been recorded for rainwater collected on Kauai during the eruption of Kilauea Iki on Hawaii in 1959–60; the steam condensate from this eruption was 1.9 *N* HCl (Murata, 1966, p. C2). Rain falling on the island of Hawaii collected by Eriksson in 1957 had a much lower pH than would be expected for an island environment.

The composition of rainwater as a chemical agent of weathering is very important, and is a factor to be considered in the chemical climate of any region. Rainwater itself is a mixed electrolyte that contains varying amounts of major and minor cations and anions. Sodium, potassium, magnesium, calcium, chloride, bicarbonate, and sulfate ions are major constituents. Ammonia and various nitrogen compounds are always present. Minor constituents include iodine, bromine, boron, iron, aluminum, and silicon. Dust particles are added locally in industrial areas, large population centers, and arid areas. The sources of these constituents are the oceans, fresh-water and saline lakes, land masses, vegetation, man-made industries, and volcanic emanations.

The amount of chemical weathering is dependent on the amount of water that is available to leach rocks and on the drainage which is determined by the porosity and the position of the water table. When leached with water, certain ions that are dissolved from the minerals in rocks are found in the drainage water that has passed through the rock. Movement of water in the zone of weathering is controlled by gravity, capillarity, humidity, and heat. There are four types of water commonly present:

1. Gravitational: percolates downward and drains away carrying soluble ions
2. Capillary: held by capillary action in pores and small spaces

TABLE 26

Chemical Composition of Rainwater in Various Localities (compiled by the author)

Locality	Rainfall, mm/year	pH	Chemical element, mg/liter							
			S	Cl	NO_3–N	NH_3–N	Na	K	Mg	Ca
Europe, north	560	5.47	1.46	3.47	0.27	0.41	2.05	0.35	0.39	1.42
Australia, southeast	590	—	tr	4.43	—	—	2.46	0.37	0.50	1.20
Australia, south	655	—	0.12	13.34	—	—	5.06	2.05	2.20	4.21
Australia, southwest	806	—	0.13	6.81	—	—	2.70	0.39	0.48	0.45
Perth, Western Australia	890	6.0	0.28	11.7	0.03	0.03	6.44	0.39	0.84	1.40
Georgetown, British Guiana	—	—	0.42	2.93	0.86	0.02	1.51	0.18	0.31	0.75
Kampala, Uganda	1300	7.8	0.6	0.9	1.7	0.63	1.7	1.7	—	0.05
Hawaii	—	5.3	0.50	6.87	0.03	0.08	3.91	0.39	0.99	0.74
Bermuda	1458	—	0.70	12.41	0.56	0.07	7.23	0.36	—	2.91
Washington, D.C.	1052	—	0.89	0.35	2.4	0.43	0.23	0.18	—	0.23
Urbana, Illinois	940	—	0.80	0.69	1.27	0.09	0.90	0.07	—	—
Tacoma, Washington	2032	—	1.13	22.58	0.99	0.05	14.30	0.59	—	0.73
Cape Hatteras, North Carolina	1370	—	0.59	6.50	1.03	0.11	4.49	0.24	—	0.44
Columbia, Missouri	1016	—	0.80	0.15	3.81	0.44	0.33	0.31	—	2.18
Grand Junction, Colorado	226	—	1.58	0.28	2.63	0.33	0.69	0.17	—	3.41
Fresno, California	240	—	0.36	0.35	2.94	2.21	0.30	1.11	—	0.37
San Diego, California	277	—	1.11	3.31	3.13	1.15	2.17	1.21	—	0.67
England, Lake District	1240	4.45	1.06	3.3	—	—	1.9	0.02	0.02	0.03
England, cities	750	4.8	6.8	10.9	—	—	—	—	—	5.8
England, country	—	4.8	0.05	5.1	—	—	3.1	0.02	0.03	0.02

3. Hygroscopic: held as thin films on grains, and especially by the fine material
4. Combined: held in chemical combination and only removed by strong heat or change of mineral into another form, e.g., gibbsite, $Al_2O_3 \cdot 3H_2O$, to boehmite, $Al_2O_3 \cdot 2H_2O$

Capillary water acts as a solvent, and both it and hygroscopic water are the medium in which ion exchange takes place. In pedochemical weathering water is the agent that causes the chemical reactions that result in the production of soil profiles. The initial composition of soil water is that of the rainwater that falls on it, but the composition is modified by the chemical elements in the hygroscopic water already there.

The pH of natural water is controlled by chemical reactions and equilibrium among the ions in solution in it. The pH of rainwater is primarily due to the absorption of CO_2, and at P_{CO_2} at 25°C, when it is saturated with CO_2, the pH is approximately 5. Such water produces carbonic acid of 0.03 *M*. Rainwater in clouds has a pH of 5.7, but this value varies according to the other cations that are present (Carroll, 1962b). Not all rainwater is saturated with CO_2; there is a delicate equilibrium between the CO_2, the H_2CO_3, and the HCO_3^- content, which can be stated thermodynamically as

$$\frac{[H^+][HCO_3^-]}{[H_2CO_3]} = K_{H_2CO_3} = 10^{-6.4} \qquad \text{at } 25°\text{C}.$$

This constant varies from 6.58 at 0°C to 6.33 at 30°C (Garrels and Christ, 1965, p. 89), which is the pH range in which weathering takes place.

All water that comes in contact with rocks is slightly acid. This is the most important fact of chemical weathering. The pH of water received by rocks from rain is changed by hydrolysis reactions (Section 8–4). If mineral grains are crushed and placed in water the pH obtained is known as the abrasion pH. The following are some abrasion pH values for common minerals: quartz, pH 6–7; feldspars, pH 8–9; amphiboles, pH 10–11; pyroxenes, pH 8–10; micas, pH 7–9; calcite and dolomite, pH 8–10; clay minerals, pH 6–7.

In pedochemical weathering of geochemically weathered rock the same initial conditions of rainwater, that is, CO_2 content, pH, and the quantity received, apply as for unweathered rocks. However, the presence of organic matter and the removal or addition of cations during soil formation cause variation in the pH of the rainwater and in its composition, and hence affect the actual chemical process of weathering. Groundwater, streams, rivers, and lakes contain the soluble products of chemical weathering. Large quantities of chemical elements are carried into the oceans by rivers. It has been estimated that in many drainage basins each square mile loses between 70 and 80 tons of soluble matter per year.

Most rocks consist of minerals that are not easily soluble, but in geologic time considerable quantities of the original rock minerals release chemical elements to circulating water. There is a general relationship between the mineral composition of rocks and that of the water in the rocks, that is, the water derived from rain penetrating the rocks. Data obtained from numerous chemical analyses of water in four different groups of rocks were assembled by Hem (1959). There are differences in composition of water draining through igneous and metamorphic rocks, resistates (sandstones), and hydrolyzates (clays, laterites, and bauxites). The kinds of water associated with these groups of rocks are shown in Figs. 21 and 22.

The water in wells in granite and in diorite in North Carolina differs in its content of chemical elements (Le Grand, 1958). Water in granite has a pH of 6.5, that in diorite, 7.1. The dissolved solids in the water in granite are 75 ppm, those in diorite, 269 ppm. The water in diorite contains many times the amount of Fe, Ca, Mg, Na, and K cations and CO_3, SO_4, Cl, F, and NO_3 anions than does the water in granite. The total hardness (as $CaCO_3$) of the water in diorite is 7 times as great as that of the water in granite.

Chemical analyses of the water in 19 springs draining laterite in central Africa (Gimbi, formerly Belgian Congo, 5°26′ S, 13°59′ E), as reported by Waegemans (1954), show the following characteristics: pH varies from 4.84 to 6.82, with an average of 5.92; SiO_2 varies from 4.9 ppm to 31.4 ppm, averaging 14.9 ppm; FeO, <1 ppm; CaO, 1.2–21.2 ppm; MgO, <1–5.2 ppm; Na_2O, 2.2–17.8 ppm; K_2O, <1–4.7 ppm; Al_2O_3, <1 ppm; Cl, <1–15.6 ppm; HCO_3, 10.3–66 ppm; PO_4, <1–11.2 ppm; NO_3, <1 ppm. The highest figures, apparently, are from water that comes from a somewhat different terrain. The springs are situated on a plateau 425–475 m above sea level; the rainfall, over 1300 mm annually (51 inches), is fairly well distributed in all months except June, July, August, and September, which constitute the dry season.

8–3. Solution

Chemical weathering depends on the reactions of silica and alumina with water and other very dilute solutions, as the major constituents of rocks are combined as aluminosilicate minerals. Silica is slightly soluble at all pH values, whereas alumina is only soluble below pH 4 and above pH 8.5. The solubility of silica is 6 ppm; quartz therefore is slightly soluble at the temperature of rock weathering. Amorphous silica is soluble to the

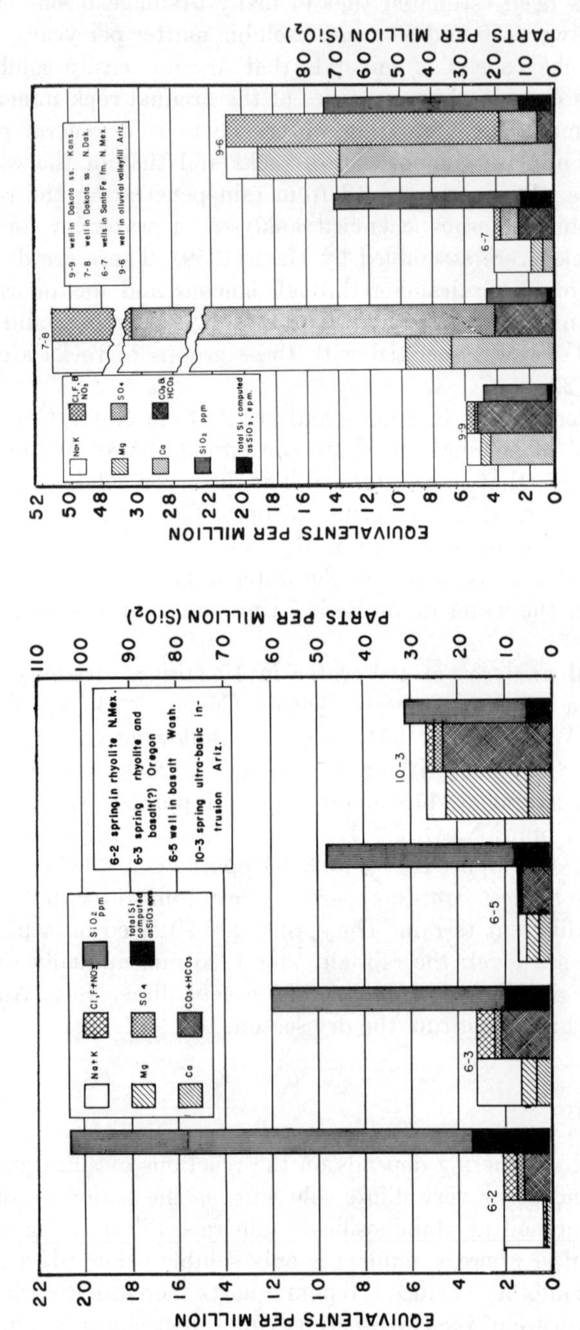

Fig. 21. Chemical composition of waters associated with igneous (left) and resistate (right) rocks (after Hem, 1959).

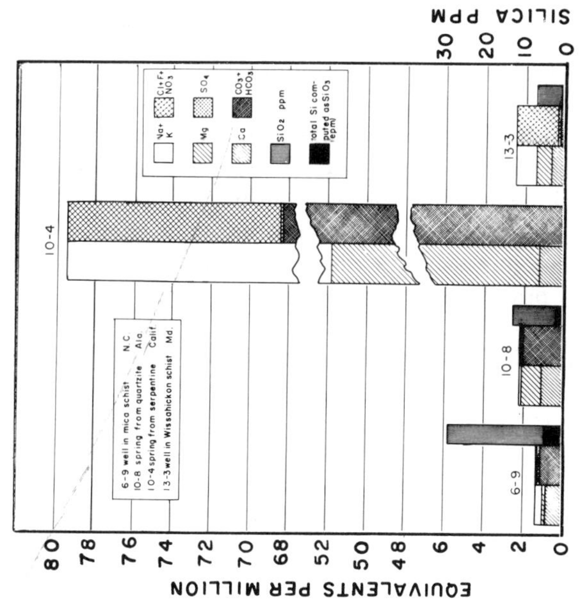

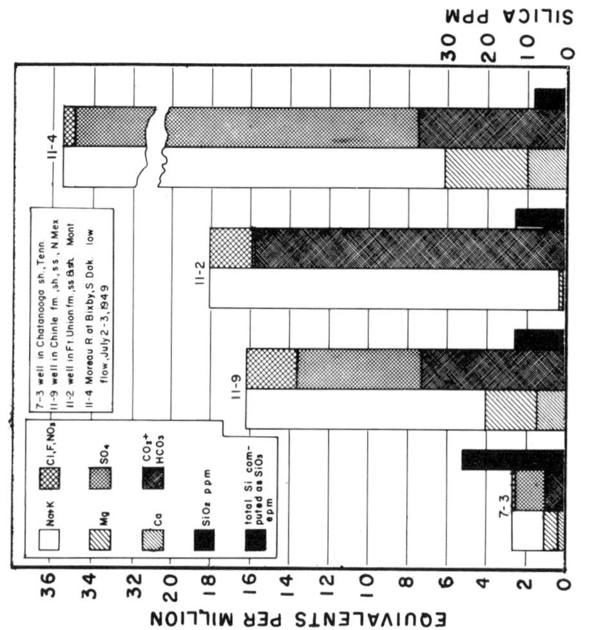

Fig. 22. Chemical composition of waters associated with hydrolyzate (left) and metamorphic (right) rocks (after Hem, 1959).

extent of 115 ppm, or nearly 20 times as soluble as quartz, but except in sandstones is not nearly as abundant as quartz. At pH values above 9 the solubility of silica increases abruptly because of the ionization of H_4SiO_4, but in weathering pH 9 is only attained during the initial alteration when feldspars become hydrolized. This pH value is attained in calcareous and alkali soils. For the most part, in geochemical and pedochemical weathering silica in small quantities is continually leached out and alumina accumulates in the clayey residuum. The behavior of silica in weathering and sedimentation has been described in detail by Siever (1962).

Alumina is practically insoluble between pH 4 and pH 8.5, that is, in the pH field of rock weathering and soil formation. Alumina is amphoteric; at low pH values it is in solution as $Al(OH)_2$, and at high pH values as $Al(OH)_4^-$ (Hem and Roberson, 1967; Hem 1969). Figure 23 shows the solubility of quartz, amorphous silica, and alumina as a function of pH.

Minerals are structural combinations of cations with oxygen. The minerals have different degrees of stability depending on their chemical constituents and their crystalline form. Slightly soluble minerals are known

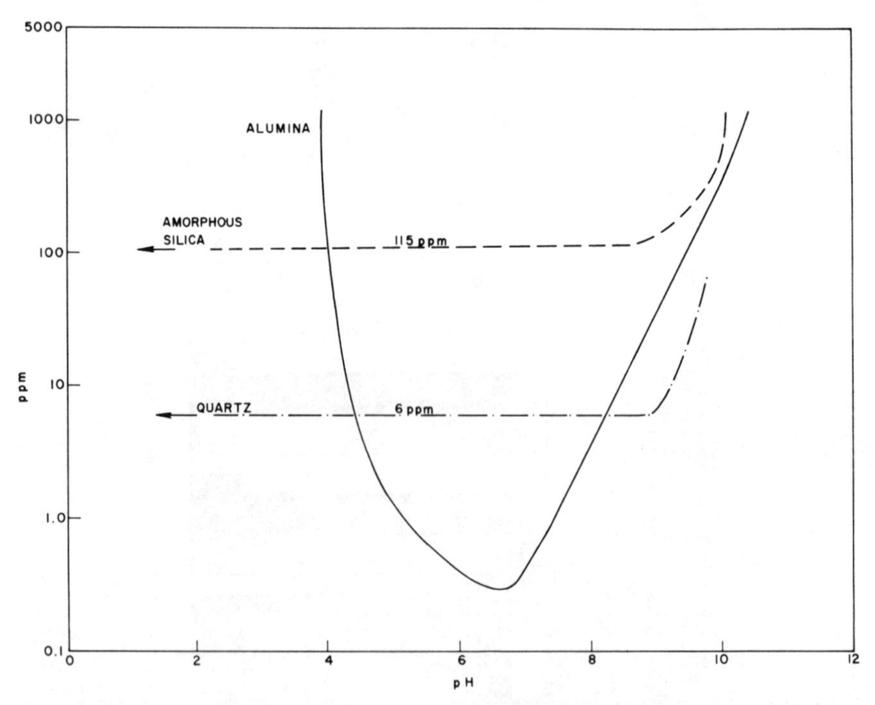

Fig. 23. Solubility of quartz, amorphous silica, and alumina as a function of pH (quartz data after Siever, 1962; amorphous silica data after Morey et al., 1964; alumina data after Magistad, 1925).

as very stable or resistant; they accumulate when present in weathering situations and, if commercially useful to man, they are known as placers (gold, tin, zircon, etc.). This accumulation is most readily recognized in mature soils that have not been eroded. Such soils may form the raw material of new sediments after transportation and deposition.

Minerals vary under acid attack. Some are insoluble, some gelatinize with acid, and some partially dissolve leaving siliceous frameworks. The insoluble or difficultly soluble minerals consist of those that are commonly present in resistates. The commonest of these is quartz; others are zircon, tourmaline, ilmenite, sillimanite, and kyanite. The minerals that gelatinize with acid are olivine, andradite, nepheline, some zeolites, and nontronite. These minerals have structures of two-dimensional or three-dimensional linkages of silicon and oxygen. The minerals that leave siliceous frameworks when attacked by acid have sheet structures consisting of silica tetrahedra and alumina octahedra, the long, continuous sheets becoming disintegrated during acid attack. Micas, chlorites, and clay minerals are of this type (Murata, 1943). These reactions form the basis of the stability series of Goldich (1938) in Section 4–3.

The behavior of minerals attacked by acid is very important in weathering. The ions that are dissolved remain in solution where they may be removed by circulating water, react with other ions or with minerals either to form new mineral combinations or to modify existing minerals, or enter into exchange reactions with other minerals. The presence of ions dissolved from minerals in a solution alters the composition of that solution and therefore its ability to react with unaltered minerals. Murata (1943; p. 26) made a fundamental observation about the difference between silica gel and the silica released by acid attack of sheet structures: the silica gel formed when a mineral gelatinizes under acid attack has a surface area of about 600 m^2/g, whereas the silica separated from glauconite (a micaceous mineral) has a surface area of about 80 m^2/g.

As has been mentioned earlier, atmospheric CO_2 is absorbed by water to form solutions of weak carbonic acid that react with minerals more strongly than pure water. This process is one of the most important whereby the surface of the lithosphere is lowered by chemical denudation. Carbon dioxide is absorbed from the atmosphere by rainwater, and by soil water from the CO_2 released by the growth of plants and the metabolic processes of microorganisms in the soil. In geochemical weathering the process is inorganic, whereas in pedochemical weathering the process is a complex of organic and inorganic reactions. Inasmuch as carbonation takes place in the presence of water it is insignificant in arid climates. The carbonation process is particularly important in the chemical winnowing of limestone and dolomite, whereby caves are formed. The process is also important in soils and its incidence forms the dividing line between the two major soil classes, the pedalfers (leached), and the pedocals (unleached), as described in Section 5–2.

In addition to causing the removal of limestone by dissolving calcium carbonate with weak acid, any detrital constituents in the limestone, such as clay, accumulate to form soils and, with time, certain pocket types of bauxite. Soils formed on limestones are of two types—the terra rossa of the Mediterranean region on limestone karst topography and the rendzina, a black soil in more humid regions where leaching, although it does occur to some extent, is not important since such limestones are only slightly porous.

The process of leaching and accumulation of clay minerals in limestones was examined by Carroll and Starkey (1959). Leaching is based on the production of $CaHCO_3$ and H_2CO_3 from limestone by leaching with carbonated water. The amount of CO_2 soluble in water under one atmosphere pressure of CO_2 at 25°C is 1.45 g/liter. This produces a weak acid, about 0.03 M. The amount of $CaCO_3$ soluble in water varies with the amount of CO_2 in solution (Garrels and Dreyer, 1952), temperature, added salt, and pH. Miller (1952) states that in equilibrium with the P_{CO_2} of the atmosphere the solubility of $CaCO_3$ is 0.86 g/liter. Carroll and Starkey (1959) showed experimentally that alumina and ferric oxide are concentrated from clay minerals in limestones by the continuous passage of surface waters through limestones and that silica will gradually be leached out.

The principal reactions are shown by these simplified equations:

$H_2O + CO_2 \rightarrow H_2CO_3 + (HCO_3)^-$—Production of carbonic acid

$CaCO_3 + H^+ + (HCO_3)^- \rightarrow Ca^{2+} + 2(HCO_3)^-$—In solution; limestone reaction

$Ca, Mg(CO_3)^2 + 2CO_2 + 2H_2O \rightarrow Ca(HCO_32_2 + Mg(HCO_3)_2$—Dolomite reaction

Calcite is more soluble than dolomite; an example is the Luray caves in Virginia, where calcite has been leached differentially from dolomite which now forms the framework of the cave. Continuous leaching is necessary in this process because the rate of solution of $CaCO_3$ decreases as the water becomes saturated with calcium and the pH increases. As CO_2 is continuously added to the water, the pH is lowered (more acid is formed) and hence more $CaCO_3$ is removed.

The effect of leaching by CO_2-enriched water, although the commonest and most frequent process of chemical denudation in all regions except the polar, is one of the most complicated in nature. Rainwater can dissolve 3.1 tons of calcium carbonate per square mile per inch of rainfall from a limestone area, provided that equilibrium in the system $CaCO_3$–CO_2–H_2O is attained and provided the $CaCO_3$, once dissolved, is not precipitated (Miller, 1952; p. 195). Under humid conditions (assuming a rainfall of 40 inches per year) the rate of lowering the land surface would be 1 ft in 44,700 years. Groundwater, because of its complex chemical composition, can dissolve more calcium carbonate than can a similar quantity of rainwater.

From the foregoing discussion it is apparent that some chemical elements, namely, Ca, Mg, Na, and K, will be removed from weathering rocks more rapidly than others, and that alumina accumulates in the clay complex. The process of element removal can be called chemical winnowing. There are only slight differences in the mobility of elements removed from different types of rocks by chemical winnowing under steady-state conditions over broad areas. Anderson and Hawkes (1958; p. 210) found that in granitic and schistose rocks from three individual drainage basins in New Hampshire, the order of mobility was

$$\mathrm{Mg} > \mathrm{Ca} > \mathrm{Na} > \mathrm{K} > \mathrm{Si} > \mathrm{Al} = \mathrm{Fe}$$

Smyth (1913) and Polynov (1937) found that the order of mobility was

$$\mathrm{Ca} > \mathrm{Na} > \mathrm{Mg} > \mathrm{K} > \mathrm{Si} > \mathrm{Al} = \mathrm{Fe}$$

Miller (1961) found the following relative mobilities of elements in small streams draining the Sangre de Cristo range in northern New Mexico:

Granite: $\mathrm{Ca} > \mathrm{Mg} > \mathrm{Na} > \mathrm{Ba} > \mathrm{K} > \mathrm{Si} > \mathrm{Fe} = \mathrm{Mn} > \mathrm{Ti} > \mathrm{Al}$

Quartzite: $\mathrm{Ca} >> \mathrm{Na} >> \mathrm{K} > \mathrm{Mg} > \mathrm{Fe} > \mathrm{Si} > \mathrm{Al}$

Sandstone: $\mathrm{Ca} >>> \mathrm{Na} > \mathrm{K} > \mathrm{Si} = \mathrm{Al}$

8–4. Hydrolysis

Chemical weathering is caused by hydrolysis of rock minerals. Hydrolysis is a reaction between a salt and water to give an acid and a base. Aluminosilicate minerals are salts of weak acids, silicic acid, and aluminum which is amphoteric, that is, aluminum can act either as a base or as an acid. Water is also amphoteric. Silica combined with oxygen and aluminum in crystalline minerals is probably more soluble than quartz, but less soluble than in amorphous form. Water with its content of cations and anions is the reagent that causes hydrolysis; the amount of H^+ (strictly, H_3O^+) dissociated in water is indicated by the pH of the water, which is the expression of the chemical environment of weathering (Section 2–1). The abrasion pH of a crushed mineral is its hydrolysis reaction. The pH value indicates whether H^+ or OH^- remains in the water in contact with the mineral. In the copper sulfate hydrolysis equation, $CuSO_4 + 2H_2O \rightarrow Cu(OH)_2 + H_2SO_4$, the pH of the water is low (H_2SO_4 is present), whereas the pH obtained with crushed feldspar is 8–9, indicating that H^+ has become absorbed, leaving OH^- in solution. Salts of strong acids and bases dissociate in water giving a pH that is almost neutral, but minerals that are constituted of strong bases (strongly ionized cations) and weak acids (weakly ionized anions), such as the silicate minerals of rocks, are of this type and generally have high abrasive indices.

The properties of both silicon and aluminum add to the complexity of the hydrolyzation of minerals. (The solubility of both silica and alumina was discussed in Section 8–3.) At low pH values silica forms monomeric H_4SiO_4, which is a weak acid and largely un-ionized below pH 9. The hydrolyzation of alumina is complex. In acid solutions, pH 2.85–4.07, a cationic form exists, $Al(OH)_2^+$, and in alkaline solutions, pH 10–12.5, an anionic form exists, $Al(OH)_4^-$, with an isoelectric point at pH 6.70 (Fig. 22, Section 8–3) (Reeseman et al., 1969).

Data concerning the solubility of alumina were reviewed by Hem and Roberson (1967) who showed that within the range of acid solubility most reactions take place very slowly, and that temperature, concentration, and addition of reagents influence the final form the aluminum hydroxide will assume. In addition, aluminum hydroxide tends to form colloidal or subcolloidal suspensions. When the pH of a solution containing Al^{+3} ions is raised, dissolved species consisting of associated Al^+: and OH^- ions are formed; when the supply of OH^- ions is sufficient, a precipitate, $Al(OH)_3$, is formed. If the solution becomes more alkaline the precipitate is redissolved. The range of pH in most weathering situations is precisely that of Fig. 23 (Section 8–3) in which aluminum exists as various kinds of hydroxides below the isoelectric point. It is true that some soils have a pH of 7–9, but these belong to the pedocal group and any aluminum in solution will be in the anionic form.

Minerals do not react in the same way as chemicals in a laboratory, and the conditions under which hydrolyzation takes place are generally stable for geologic time (the weathering environment, Sections 2–1 and 2–2). This stability of reaction leads to a continual change in the same direction of the silicate minerals. However, minerals react at different rates, so that a weathering series, or stability series, as formulated by Goldich (1938) results.

Investigations by Nash and Marshall at the University of Missouri and by Garrels and Howard at Harvard have helped elucidate the process of hydrolysis from the mineralogical point of view. Feldspar surfaces in contact with water react to form a thin film in which H^+ ions have replaced

K^+ ions and the mineral structure becomes slightly disorganized. In the plagioclase feldspars (Na–Ca), the cation that is present in smaller amount is replaced by H^+ ions first, i.e., if Na $>>$ Ca, then Ca is removed preferentially. Biotite mica loses its cohesion easily by the replacement of the interlayer K^+ by H^+, or by Mg^{+2} (forming vermiculite). Muscovite mica is much more stable and hence it is found in soils and sedimentary rocks.

The atomic structure of feldspar helps to explain hydrolysis during leaching by water. Feldspars consist of frameworks of linked silica and alumina tetrahedra forming a three-dimensional chain structure, the links consisting of four rings of tetrahedra—three silicon to one aluminum. The cations of the structure are situated in the interstices of the negatively charged framework. In orthoclase feldspars the cations, K and Ba, are large (K, 1.33 Å; Ba, 1.34 Å, ionic diameter); in plagioclase feldspars the cations are small (Na, 0.97 Å; Ca, 0.99 Å, ionic diameter). In hydrolysis the silica tetrahedra form dilute silicic acid (H_4SiO_4). The alumina tetrahedra form $Al(OH)_2^+$ at pH < 4, and an amorphous precipitate, $Al(OH)_3$, at pH > 4 (Fig. 23). If the solution containing the amorphous precipitate becomes more alkaline, the precipitate crystallizes as bayerite. Below pH 7 the dissolved aluminum species consists of octahedral units in which aluminum is surrounded by six water molecules or hydroxide ions. Single units such as $Al(OH_2)^{+3}$ and $AlOH(OH)_5^{+2}$ are most abundant below pH 5 where the molar ratio of the combined hydroxide to total dissolved aluminum is low. When the molar ratio is greater than 1.0, polymerization of the octahedral units occurs. When the molar ratio is between 2 and 3, solutions aged for 10 days or more contain colloidal particles between 0.10 and 0.45 μ. Particles of larger diameter than 0.10 μ have the x-ray diffraction pattern of gibbsite. Aging increases the size and orderliness of the polymeric aggregates and is accompanied by a decrease in pH of the solution (Hem and Roberson, 1967; p. A1).

Geologists and soil scientists long appreciated that hydrolysis was the reaction whereby clays, laterites, and bauxites accumulated, but no detailed chemical investigations had been made to describe the complex chemical reactions that took place. Hydrolysis of silicate minerals in geochemical systems determined by the chemical environment of a region proceeds in a stepwise manner with the reaction of water with silica and alumina frameworks of the silicate minerals. With time the end product appears as though only simple reactions of water with minerals had taken place. One early description of hydrolysis of a basic rock to primary gibbsite in British Guiana by Harrison (1934; p. 37) showed that gibbsite formed directly from plagioclase feldspars. The leaching conditions at this locality were probably maximum—117 inches of rain annually, and very few dry periods. Harrison stated that only eight short dry periods occurred during the 10 years he was stationed there, and that from April, 1909 to January, 1912 rain was continuous.

Hydrolysis of the minerals in basalts in the high rainfall areas of the Hawaiian Islands produces desilication and large amounts of halloysite and gibbsite (Bates, 1962). The first step in the hydrolysis of feldspars is the formation of halloysite (a clay mineral of similar composition to kaolinite, but with additional water), which produces gibbsite when desilicated, apparently by the chemical reactions described by Hem and Roberson (1967). Prior to aging, the aluminum precipitates from desilication produce allophane, which is amorphous to x-rays and has little or no crystalline structure. Another manifestiation of desilication is the movement of amorphous aluminum precipitates—gels, which are orange colored due to iron oxide. These gels may coat rocks, accumulate in stagnant water, or trickle down stream banks (Bates, 1960).

The hydrolysis of feldspars under strong leaching conditions has been observed in many places. One occurrence has been described in detail by Millot and Bonifas (1955) in the desilication of dunite at Conkery, French West Africa. The weathered dunite has a mottled appearance which is described as *pain d'épice.*

The movement of alumina gel and its subsequent crystallization as nordstrandite ($Al(OH)_3$) in some limestones of Guam has been described by Hathaway and Schlanger (1965). The alumina gel from which nordstrandite crystallized apparently migrated to solution voids in the limestones, but did not replace calcite. The aluminum hydroxide precipitates formed between pH 7.5 and 9.5 are initially amorphous, but on aging crystallized as bayerite (Hem and Roberson, 1967). The pH of the limestone environment is between 8 and 9. The alumina gel probably came from the desilication of halloysite which is detrital in the limestones; movement of this gel into the voids in the limestones provided the material to crystallize as nordstrandite (Hathaway and Carroll, 1964; p. D44). It is probable that nordstrandite will be identified in the products of desilication of basic igneous rocks in regions where active hydrolysis is taking place. One such occurrence is in the soils of Borneo (Wall et al., 1962).

In *hydration* mineral grains adsorb water. Often the first uptake of water is not connected with changes in the mineral structure. The adsorptive capacity increases with increasing ionic charge. Some elements readily become hydrated, such as ferric oxide, whereas others never do, such as silicon in quartz. Hydration generally accompanies carbonation (Chapter 7). A common example of hydration is the adherence of water to ferric oxide:

$$2Fe_2O_3 + 3H_2O \rightarrow 2Fe_2O_3 \cdot 3H_2O$$

which is an amorphous, hydrated iron oxide that in time will crystallize as goethite, $Fe_2O_3 \cdot H_2O$, an orthorhombic mineral.

8–5. Ion Exchange

The most important process in the chemical weathering of rocks is ion exchange. Theoretically, this is a simple reaction between ions in a solution and those held by mineral grains. Ion exchange is defined as the amount of exchangeable cations, expressed as milliequivalents per 100 grams of material, determined at pH 7 under experimental conditions. The reaction is of the type

$$\text{Na-clay} + H^+ \rightarrow \text{H-clay} + Na^+$$

However, the mechanisms of the exchange of ions are not simple (Carroll, 1959b).

Considered as a part of the weathering process, cation exchange in rocks, weathered rocks, soils, and unconsolidated sediments is due mainly to the kinds and quantities of minerals in the silt (2–62 μ) and clay ($<2\ \mu$) particle sizes. The clay minerals are chiefly responsible for the exchange capacity of weathering rocks and soils. Table 27 gives an outline of the essential chemical and structural features of the principal clay minerals. The principal kinds of clay minerals are, structurally, sheets of combined silica tetrahedra and alumina octahedra. The clay mineral acts as a substrate for the reactive ion, i.e., the ion that is exchangeable, The exchangeable ion is held to the mineral surface because within the clay mineral itself there is a charge deficiency and the mineral is not electrically neutral. A cation that becomes adherent to the clay surface neutralizes the negative charge that has developed within the clay mineral because of isomorphic replacements within the clay structure or crystal framework. A commonly found replacement is that of Si^{+4} by Al^{+3}. Such a replacement is found in the mineral montmorillonite. The structure of clay minerals has been described by Grim (1953) and Brown (1961).

Each kind of clay mineral has a different charge deficiency that can be expressed as its ion-exchange capacity. In addition, nonclay minerals also have a small ion-exchange capacity caused by unsatisfied valences at the surface. Humus, organic material, and humus derivatives have a high exchange capacity. Representative figures for the ion-exchange capacity of some minerals and other materials are given in Table 28.

There are four kinds of cation exchange that are important in weathering:

1. Exchange between cations in a solution in contact with a mineral and the exchangeable cations of the mineral. The exchange takes

place in water surrounding the mineral grain. The groundwater is the electrolyte carrying the cations in geochemical weathering; the soil water is responsible for reactions in soils.
2. Contact exchange, in which cations in one mineral are exchanged for those in another. An example of this kind of exchange is the reaction of an exchange resin on a mineral, such as the removal of calcium from limestone by contact with an H-resin; or the H-form of clay reacting with mineral grains (Graham, 1941).
3. Exchange due to the association of an organic compound with a mineral, e.g., amines with montmorillonite, whereby the original exchangeable cations are replaced by amine molecules between the layers.
4. Adsorption of ions from mineral particles by the root hairs of plants (Jenny, 1951). The ion swarms of the root hairs and the soil particles intermingle, and ions are adsorbed because certain cells of the root hairs act as exchange sites.

TABLE 27

Chemical and Structural Features of Clay Minerals

Group and formula	Structure factors			Ion-exchange capacity, meq/100 g
	Silica tetrahedra	Alumina octahedra	Interlayer cation	
Kaolinite $Al_4Si_4O_{10}$	1	1	none	3–15
Mica $H_2KAl_3(SiO_4)_3$	2	1	K	1–5
Illite (hydrous mica)	2	1	K	10–40
Mica + mixed-layer glauconite $K(Al, Fe, Mg, Ca)_2(Si, Al)_4O_{10}(OH)_2$	2	1	K, Mg, Ca	
Montmorillonite $(Al, Mg, Fe)_2(Si, Al)_4O_{10}$	2	1	H, Mg, Na, Ca	80–100
Chlorite $(Mg, Fe, Al):(Si, Al)_4O_{10}(OH)_8$	2	1	none	1–5
Vermiculite $(Mg)_3(Si, Al)_4O_{10}\cdot H_2O$	2	1	Mg, H_2O	120

Data are for well-crystallized minerals and not for mixed-layer structures, except ion-exchange capacity for illite.

TABLE 28

Ion-Exchange Capacity of Representative Materials

Mineral	Ion-exchange capacity, meq/100 g	Remarks
Kaolinite	3–15	Highest figures for finer grain size and disordered structure
Halloysite ($2H_2O$)	5–10	Increases with acid treatment
Halloysite ($4H_2O$)	40–50	
Montmorillonite group	36–100	Most members of the group are in the range 70–100; saponite, sauconite, and stevensite have much lower figures
Illite	10–40	
Vermiculite	100–150	Varies with interlayer cation
Chlorite	4–47	Varies with particle size, composition, and crystallinity
Glauconite	11–20	
Palygorskite group	20–30	
Allophane	about 70	Varies with degree of crystallinity
Feldspathoids		Varies with differences in structure
Zeolites	100–300	Varies for different zeolites
Talc	0.2	
Pyrophyllite	4.0	
Feldspar	<1	Depends on grain size
Quartz	<1	Depends on grain size
Pumiceous tuff (Tsherege member of Bandolier tuff)	1.2	Mesh, <230
Basalt (Snake River)	0.5	Whole fragments >2 mm; increases with reduction of grain size
Barstow formation (Miocene ?, Mojave Desert)	32 (whole rocks), 49 (>2 μ)	Average of six samples containing heulandite or phillipsite, and mordenite or analcite
Organic matter in soils	130–350	Varies according to climate and vegetation

Of these, the first two types of exchange are primarily effective in geochemical weathering, whereas the last two listed are necessary processes in pedochemical weathering and the growth of vegetation.

In Table 28 a range of exchange capacity is given for many minerals. The principal reason for this range is that the crystallinity of the individual minerals varies from a perfect crystal structure to a poorly crystalline and disordered structure. The hydrous micas and illites illustrate this. The more perfect the structure, the more nearly is the crystal electrically neutral, except for isomorphic replacement either in the alumina octahedra or silica tetrahedra. In mica, for example, K^+ bonds the layers firmly together; but in illites some K^+ has been removed and the layers are not firmly bonded together, additional surfaces are exposed, and the exchange capacity is greater than in the mica structure. Kaolinite and disordered kaolinite provide another example. An x-ray diffractogram shows the variation in crystallinity in the variation of the main (basal) *d* spacing in oriented mounts of the mineral.

The net effect of these exchange processes is to remove or to add cations to mineral grains, thereby effecting a change in composition. The exchange that takes place is partly governed by the composition and pH of the water that is in contact with the mineral grains. This applies particularly to the first process listed above, which is of extreme importance and has resulted in the establishment of the U.S. Salinity Laboratory to study these reactions. In general, water with a pH up to 5 will replace metal cations with H^+, but water that is neutral (pH 7) will allow the metal cations to remain in the exchange positions, or will replace H^+ with metal cations, depending on the composition of the water.

The common metal cations, Na^+, K^+, Ca^{+2}, and Mg^{+2}, are readily exchangeable, and Zn^{+2}, Cu^{+2}, Co^{+2}, Mn^{+2}, and others are also known to be exchangeable. Divalent cations are more strongly held, and therefore less easily exchanged, than univalent cations; but part of the complexity of cation exchange is that the cations all differ in their relative ease of replacement, the order of which is $Li < Na < H < K < Mg < Ca$. However, H^+ often seems to be more tightly bonded to the exchange sites in minerals than Na^+ or K^+. The calcium ion is the dominant cation in poorly drained soils or those in dry climates (pedocals).

At low pH values (up to pH 4) many minerals, particularly clay minerals, contain readily replaceable aluminum that can be removed with a neutral salt. Paver and Marshall (1934) consider that this Al^{+3} ion acts as a replaceable base that augments the so-called exchangeable acidity. In clay minerals Al^{+3} is part of the octahedral layer, and if removed the mineral structure breaks down. This mechanism is an important part of mineral alteration under acid conditions.

In pedochemical weathering, the total reaction of a soil (generally measured as the pH of a 1:10 soil–water mixture) forms the basis of a great deal of our knowledge about the use and geographical distribution of the different kinds of soils. The determination of soil pH is routine in all places where soils are investigated scientifically. Such pH values give an overall picture of the base status of a soil, that is, the percentage base saturation of the clay complex or the amount of cations that the clay holds in its exchange positions. For example, if 50% of the possible exchange sites are filled with cations other than H^+, then the base saturation is said to be 50%. A 50% base-saturated clay will have a higher pH value than if it were base unsaturated, implying that H^+ ions occupy most of the exchange positions.

Two other aspects of cation exchange are of interest and importance in weathering. The first of these is fixation of certain cations in micaceous structures and the second is the clogging of exchange sites of minerals by iron oxides.

The structural sheets of micaceous minerals are held together by K^+ ions. By the processes of chemical weathering, these ions are leached out and replaced by hydrogen or by hydronium ions, and the structure becomes a degraded mica, generally called an illite. An excess negative charge develops. Metal cations can enter the degraded structure, if present in the surrounding solutions, and the mica is regenerated. Rolfe and Jeffries (1952) were the first to point out that this fixation of potassium takes place on a large scale. It had been known for a long time by agriculturalists, and the information was reviewed by Reitemeier and others in 1951. The importance of the fixing power of hydrous micas arises from the fact that they are the predominant clay minerals in soils derived from many rocks in the temperate regions that have not been subjected to long periods of soil-forming processes. Fixation of NH_4^+ ions by soil illites and vermiculites is of interest because ammonia is produced by the metabolic processes of microorganisms in the nitrogen cycle in soils. The ammonia fixation amounts to about 1 meq/100 g in moist surface soils, but is higher, nearly up to 4 meq/100 g, in subsoils. Ammonia fixation was reported in soils from British Honduras, Trinidad, and British Guiana by Rodrigues (1954), and may be a common feature in tropical soils.

The clogging effect of substances that cover the exchange sites on flat surfaces, edges, and between the layers in such minerals as montmorillonite and the micaceous minerals (mica and illite), reduces the exchange capacity of these minerals. This phenomenon is largely found in soil clays, and not in the clays formed by geochemical weathering. The iron oxide of soils, often called the free iron oxide because it is uncombined in the structure, is largely associated with the clay minerals. Fripiat and Gastuche (1952) have studied the free iron oxide that adheres to the external surfaces of kaolinite that have OH and O ions exposed. If the kaolinite has Ca^{+2}, Mg^{+2}, Na^+, or K^+ ions in the exchange positions, the structure of the kaolinite–iron oxide complex is compact, ordered, and nonporous. Saturation of the surface is produced very rapidly and excess iron oxide forms particles of pure oxide. If, however, the kaolinite is in the H-form, the resulting complex is porous and has a disordered structure. Where iron oxide is present the exchange capacity is lowered, but iron oxide facilitates the adsorption of phosphate ions and the formation of iron phosphate. Fripiat and Gastuche obtained the above results experimentally. In natural situations, the kaolinite is most often in the H-form so that the iron oxide is porous and readily removed.

The discussion of ion exchange given above has been principally concerned with cations. Anion exchange also occurs but it is complicated and not yet as completely known or documented as is cation exchange (Carroll, 1959b).

Part of the known ion-exchange capacity in soils is due to the clay minerals in both the clay ($<2\ \mu$) and the silt (2–62 μ) fractions and part is due to the presence of organic matter. Hosking (1948) showed that the exchange capacity of organic matter varies with the type of soil. He found that this exchange capacity ranges from a mean of 170 meq/100 g to a mean of 360 meq/100 g. The composition of the organic matter in a soil varies according to climatic conditions as shown by Kononova (1961) and ranges in composition from humic to fulvic acid. The cation-exchange capacity of humic acid is given by Kononova (1961, p. 55): chernozem, 474.5 meq/100 g; podzolic soil, 345.2 meq/100 g; dark chestnut soil, 483.3 meq/100 g; solonetz, 430.0 meq/100 g. Kononova states that the more alkaline the soil reaction the greater the number of functional groups of humic acid that enter into exchange reactions. Hence, the higher figures for the exchange capacity of the humic acid in the pedocal groups than in the pedalfer groups of soil.

8–6. Oxidation and Reduction

Chemical elements occur in oxidation states that are in equilibrium with their environment. The oxidation potential is low in igneous and metamorphic rocks, and in black and gray shales containing pyrite; it is high in sandstones, limestones, and red shales. Fresh and sea water are also oxidizing. The oxidation potential is measured in mV and compared to that of hydrogen, $E_H{}^\circ = 0$. A scale extends in both the positive and the negative direction from zero. Positive readings are oxidizing and negative readings are reducing. The range of readings for environments on the earth's surface is shown in Fig. 2 (Section 2–1).

Oxidation of a substance occurs when it, or one of its constituent atoms, loses an electron (e^-). Oxidation in weathering occurs by the combination of elemental oxygen with the weathering substance, that is, the adjustment of chemical elements to the oxidizing environment when these elements had previously been in a reducing environment. The oxidizing substance loses electrons to the oxygen that has become ionic. Not all elements change readily from a reduced to an oxidized state.

One of the readily oxidizable elements is iron, and the most easily recognizable first alteration in weathering is the oxidation of iron, from the ferrous state to the ferric state. In thin sections of rocks this first stage of alteration can be recognized by the presence of iron staining and of stringers of ferric oxide along cracks and between mineral grains. Keller (1957) gives the following description of oxidation:

> "Oxidation of minerals by gaseous oxygen occurs probably entirely by the intermediary of water, which may be present in quantity ranging from films of moisture to complete immersion. . . . When gaseous oxygen dissolves in water, the solution becomes reactive with an oxidizing potential which is dependent on the partial pressure of the gaseous oxygen and the acidity of the solution. Water at pH 7 exposed to air has an oxidizing potential of 810 mV. This potential is well above that necessary to oxidize ferrous to ferric iron in typical occurrences near the earth's surface. Many of the reactions in weathering and diagenesis are oxidation–reduction in type."

Water is the intermediary agent by which oxygen operates as an oxidant. Keller (1954) showed that the energy liberated upon the oxidation of iron is quite large. Iron in minerals is very susceptible to oxidation. When the ferrous which links silica tetrahedrons in a silicate structure reacts with oxygen to form ferric oxide, the silicate structure falls apart and alteration begins. This reaction can be formulated as follows:

$$4FeSiO_3 + O_2 \rightarrow 2Fe_2O_3 + 4SiO_2 + 512 \text{ kcal (liberated)}$$

The oxidation of iron is a very common and easily recognized feature. It is well known that a capping or gossan (iron hat) of iron oxide is an indication of an oxidized sulfide body which may contain concentrations of economically useful minerals. The various stages of the oxidation of a sulfide, for example pyrite, FeS_2, are given in detail by Keller (1957). The oxidation of minerals containing cations of economically useful metals is more spectacular than the oxidation of iron in common igneous and metamorphic rocks. For example, oxidized copper ore manifests itself in the minerals malachite (green) and azurite (blue). An interesting series of minerals has developed as the weathering products of uranium ores on the Colorado Plateau (Garrels, 1953). The reactions that brought about these changes have been described thermodynamically by Garrels (1954). In general, oxidation produces new kinds of minerals from those that were present in the unoxidized ore.

Most elements that we see in the weathering crust of the earth are in the oxidized state because the environment is oxidizing. Those not oxidized are generally either in their normal oxidation state or very resistant to oxidation. For example, the mineral magnetite remains unchanged for considerable lengths of time in certain environments.

Reduction cannot be separated from oxidation for both reactions are governed by the redox potential of the system.

Figure 2 (Section 2–1) shows the limits of this field as found in natural environments; it is within this field that we have to predict the reactions that can occur with the common elements. Some elements are, in their nature, unreactive (e.g., aluminum), but others are altered by the products of alteration of other minerals. For example, sulfuric acid, H_2SO_4, is a by-product of the oxidation of pyrite, and it will react with any minerals that are attacked by acid. Conversely, having identified certain minerals in soils or sediments, we are able to state the approximate pH–E_h conditions under which they formed and under which they remain stable and do not change their identity.

As soon as oxygen is present, geochemical weathering can occur, and any changeable minerals will oxidize. In pedochemical weathering, although most elements are in the oxidized state, conditions may arise where oxygen is excluded and the E_h is consequently lowered. Iron and manganese may thus be reduced in soils and mobilized. Baas–Becking et al. (1960) found that the pH and E_h of soils were:

	pH	E_h
Normal soils	2.8–10.4	+400 to +800 mV
Wet soils	3.7–8.5	+100 to +400 mV
Waterlogged soils	5.0–8.0	−350 to +200 mV

Waterlogged soils are oxygen-poor, anaerobic bacteria flourish, and ferric iron is reduced to ferrous iron which can be moved in solution. A Russian term, glei or gley, is used for such reduced soils that are characterized by a pale gray or bluish clay at the base of the B horizon. They have an impeded drainage. Iron mottles develop in reduced soils if conditions change by lowering of the water table which induces air to enter the soil. The reduced iron then oxidizes. Many profiles of laterites show this mottling. Starkey and Wight (1945) showed that there was a seasonal variation in the E_h of waterlogged soils (Fig. 24).

It has been observed that during the decomposition of organic matter (oxidation) in most soils the E_h remains above +400 mV, but in similar soils when waterlogged, the E_h drops to below +200 mV. Seasonal changes in the level of the water table cause variations in soil aeration, which in turn change conditions for the growth of different kinds of bacteria and other microorganisms, and hence cause changes in E_h. Recent investigations indicate that reduction plays an important part in the development of many soil profiles. Rice-paddy soils, for example, are reducing.

The equilibrium conditions for the various species of iron, manganese, and sulfur are described in detail in a number of recent publications by

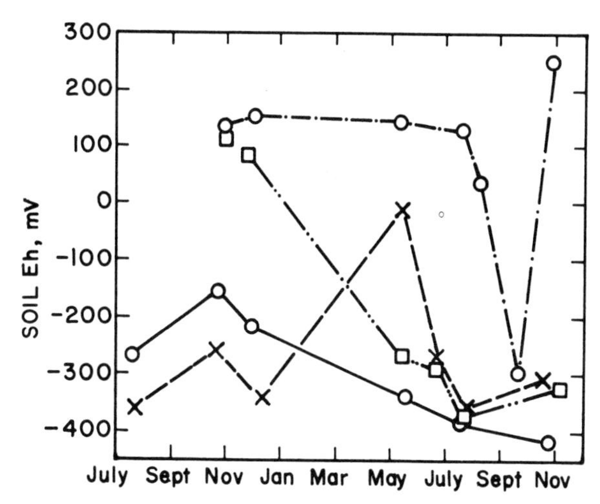

Fig. 24. Seasonal variation in soil E_h for 5 samples of soils during a two-year period (after Starkey and Wight, 1945).

John D. Hem (1959, 1960, 1963). Titanium, too, can be reduced from Ti^{+4} to Ti^{+3} when the E_h of the environment is lowered (Carroll, 1960). Removal of the iron oxides adhering to the outside of clay particles takes place in an aqueous environment with an E_h of +180 mV to +230 mV (Carroll, 1958). Thus, only a slight lowering of the E_h is required to change Fe^{+3} to Fe^{+2} and bleach the clay in soils. The reduction of Fe^{+2} is a side effect of bacterial metabolism in a closed anaerobic environment. It can readily be demonstrated in a laboratory.

Chapter 9

BIOLOGICAL ACTIVITY IN WEATHERING

The climate of any region conditions the flora that can grow there; the growth and decay of the flora provide raw organic matter which, through the action of the microflora and chemical processes, produces the soil humus. The effect of the flora and its products reacting with the mineral particles of weathered rocks changes geochemical weathering into pedochemical weathering and affects the consequent development of soils.

9–1. Role of Plants and Animals

A glance at Köppen's map (Fig. 3, Section 2–2) shows the different kinds of climates inhabited by plants and their accompanying macrofauna and microfauna. A generalized interpretation of the worldwide control of organic matter by climate is given in Fig. 25.

When plants start to grow on geochemically weathered rocks, pedochemical weathering begins and a soil is produced in time by a combination of numerous chemical and physical processes. In addition to the macroflora, trees, shrubs, grasses, and herbs which are readily seen, there is a large and very important microflora that consists largely of bacteria.

Plants produce the bulk of the organic matter in soils. The leaf litter is in the form of mor (raw humus), and mor is altered to mull (true humus). In addition to producing raw organic matter in soils, plants are important in weathering for several reasons. Both microflora and macroflora are important in changing weathering rock material. Macroflora produces the bulk of the organic matter, but this raw organic matter could not be changed into humus without the work of the microflora. Living plants and animals in the weathering material are known collectively as the biota.

Very considerable amounts of organic matter are produced each year by vegetation. Jenny (1941) assembled data on the average amounts produced by typical vegetation (given in Table 29).

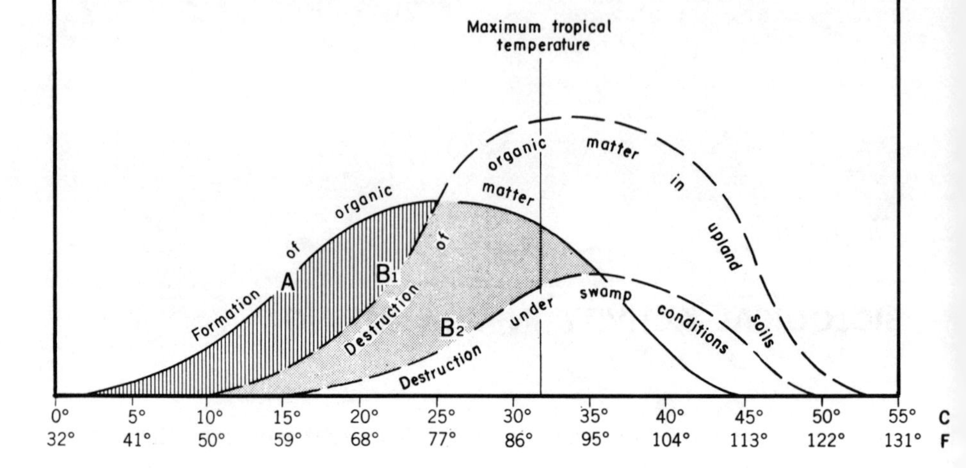

Fig. 25. Formation and destruction of organic matter in climates between the equator and the poles (after Gordon et al., 1958).

TABLE 29
Amount of Organic Matter Produced Annually by Various Kinds of Vegetation (after Jenny, 1941)

Type of vegetation	Annual production, tons per acre
Alpine meadows	0.22–0.40
Short-grass prairie	0.71
Mixed tall-grass prairie	0.85–1.73
Average forest (leaves, wood, twigs)	2.67
wood	1.42
Beech	
leaves	1.47
wood	1.42
Pine, leaves	1.42
White pine needles	2.09
Beech–birch–aspen forest (leaf litter)	2.85
Tropical primeval forest (leaves, trunks, roots)	11.1
Tropical legumes	24.4
Tropical savannah	13.3
Monsoon forest	22.2
Tropical rainforest	45–90

The macroflora provides a continuous supply of organic matter to the weathering rock debris. Trees in forested areas provide a forest floor covered with leaf litter which, if undisturbed, establishes an equilibrium between annual increment of raw humus and its destruction by microorganisms. Other types of vegetation also establish this equilibrium, but it is not as easily recognized. Thus there is an organic matter equilibrium established for every vegetation type and every climatic and microclimatic weathering situation. The organic matter is part of the steady-state condition.

The effect of the macroflora can be summarized as follows:

1. Utilizes and circulates water and contained chemical elements from the deeper layers of soil and weathered rock to the surface.
2. The larger plants provide leaf litter that is the principal ingredient of humus.
3. There is a concentration of some elements in the leaves and woody parts of plants and these elements are added to the surface soil with the leaves. The amount of fresh organic matter added to the surface has been estimated by Vageler (1913) to be 100–200 tons per hectare per year in tropical forests, 50 tons in monsoon forests, and 30 tons in savannah woodland. There is much less in colder climates.
4. The organic matter of leaves and stems provide, when leached with water, acids and other compounds that dissolve chemical elements from minerals in rocks.
5. Roots have an acid reaction that stimulates rock decomposition.
6. Roots provide channels through weathering material for the circulation of water and air.
7. Roots add CO_2 to the soil air and water, thereby increasing the production of carbonic acid which lowers the pH of circulating water, giving it increased power to dissolve minerals.
8. Root cells enter into exchange reactions with metallic cations present either in the soil water or through contact exchange.
9. Large roots penetrate the joints and cracks in solid rocks and split them into smaller pieces.

The microflora that lives in the weathering material is both numerous and varied in composition. According to Millar and Turk (1951) the top 7-inch layer of a good soil in the temperate regions has a live weight of soil biota of 2800–4200 lb per acre. This biota, except for earthworms, is soil flora: it consists of 700 lb of bacteria, 1000–1500 lb of fungi, 700 lb of actinomycetes, about 2200 lb of algae, 200–300 lb of protozoa, and 200–1000 lb of earthworms. This biota averages about 500 lb per inch thickness of soil, but the amount and composition of it vary with climate, soil use, addition of fertilizers and organic matter, and other factors. The different groups in the microflora utilize the carbon and nitrogen compounds of dead plants and animals for their nutrition and in so doing produce humus. They use oxygen in the soil air and increase the CO_2 content of soil and weathering rock materials. Algae are important colonizers of barren land, e.g., new falls of volcanic ash, coral debris on atolls, burnt land, and saline lake beds.

Factors affecting the abundance of soil microorganisms have been summarized by Corbet (1933) as follows:

1. Temperature—optimum temperature is between 25 and 37°C, but many species can exist between 10 and 45°C.
2. Moisture—the optimum for most aerobic species is between 50 and 70 percent of the saturation value. This is also the optimum value for growth of vegetation.
3. Vegetation—the relation between macroflora and microflora is indirect for most species.
4. Depth—the concentration of flora varies inversely with the depth below the surface, but practically no organisms are found below the level to which soil humic matter has penetrated.
5. Mineral content—this factor is of very little importance as only very small quantities are required.

The microfauna exists exclusively in the upper weathered part of the soil. The macrofauna can penetrate the weathered rock and act as a mechanical agent. The fauna of the weathering crust or soil consists of three groups of animals: (1) the microfauna proper, organisms that are too small to affect the pore size of the soil and live mainly in the soil-water films (protozoa, rotifers, and nematodes) and the microarthropods (mites and springtails) that live in the air spaces; (2) larger animals such as earthworms, slugs, snails, ants, termites, and others; (3) free-moving animals such as moles, rabbits, gophers, and others which have very little effect on soil development.

The first group has no effect on the grain-size distribution or crumb structure of the soil material. The members of the second group comminute the organic material and mix it with inorganic soil constituents, and also move the material. Many of these animals make burrows and thereby increase the circulation of both air and water in the soil. Members of the third group operate on a larger scale and are responsible for moving large quantities of material, thereby mixing the A and B horizons of soils. They also increase air and water circulation and cause the removal of material by running water.

In general the soil invertebrates need fairly well-aerated soil for active growth and they are therefore concentrated in the top two inches of soil. The numbers of animals present are dependent on the food supply of living or dead plant tissues and of the soil microflora. The net effect of the metabolism of the soil invertebrates is to oxidize the carbon in organic matter to CO_2 and change the complexly held nitrogen into simpler forms, thus decreasing the carbon–nitrogen ratio.

There is little information about the soil fauna of tropical areas in which the activity of the various groups is probably high. In Indonesia, Dammerman (1925) found that the number of different species in the soil fauna varied with the height above sea level. On Mount Geden the average number of species found in debris on the surface of the soil was 53/m^2 at 4500–5000 ft, 30/m^2 at 6500–8000 ft, and 11/m^2 at 9800 ft.

Both earthworms and termites are important in the Nigerian soils described by Nye (1955). The average annual rainfall is 48.4 inches and the average annual temperature is 80°F, with a relative humidity of 78% in the wet and 49% in the dry season. Earthworms deposit 20 tons of castings per acre per year under forest. The casts are richer in organic matter and exchangeable Ca and Mg, and have a higher pH, exchange capacity, and percentage base saturation than the top 6 inches of soil. Termites and other ants deposit at least one-half ton of earth per acre each year. The ants select certain sizes of grains for building mounds and thereby modify the surface soil. The earth is mainly taken from a depth of 12–30 inches. Organic carbon, pH, exchangeable Ca and Mg, and base saturation are higher in the mounds than in the surrounding soil. The activity of earthworms and ants has also been noted in many other places.

9–2. Ecology and Climatic Pattern

It is well known that certain plants and animals are adapted to living together under definite climatic conditions. They form flora–fauna associations. For example, the Everglades of Florida support flora and fauna peculiar to it, as do the Rocky Mountains, the Coastal Ranges of northern California, the Amazon Basin, central Africa, the steppes of Asia, the Mojave Desert, and hundreds of other areas that are recognizable by everyone. Ecology is the study of the relationship between climatic habitat and the flora and fauna. Habitat is affected primarily by the climate as a whole, and locally by the microclimate. Vegetative formations are an easily recognized expression of the steady state of the chemical climate of the region. Variations may be found in the ecology because of microclimate (as noted above) and because of the nature of the rocks present in an area. A major difference generally exists between the flora growing on granitic rocks and that growing on basic rocks. In addition, plants may be indicators of the presence of certain chemical elements released from weathering minerals, as, for example, uranium. Also, some chemical elements may be toxic to plants, an instance being chromium; "barrens" develop in areas where rocks supply chromium to the soil.

9–3. Organic Matter and the Formation of Humus

The vegetative cover that naturally clothes the face of the lithosphere, or that which is introduced by man, produces organic matter that forms part of the weathering process. This organic matter is at first raw material and is sometimes called raw humus. After decomposition by the microorganisms present in the surface material, this raw humus is changed into humus, which is a complex of organic substances derived from the decomposition of plant material, metabolic products of microorganisms using organic residues as a source of food, and products that are synthesized within the organic matter. Humus is the name given to organic material that has been decomposed by microorganisms. Humus is often referred to as soil organic matter. It consists of a mixture of numerous organic substances that have not yet been completely analyzed, largely because of the difficulty of isolating the components by chemical means. These substances can be called humic substances. The humus content of a soil is a measure of its fertility for two reasons: it supplies the nitrogen necessary for plant growth, and it causes aggregation of mineral grains into organomineral crumbs thereby promoting aeration, the circulation of water, and the growth of bacteria and other microorganisms in the soil. The carbon–nitrogen ratio is a measure of the change from raw organic matter to humus, and is an important attribute of a soil that is to be used for agriculture. The amount of organic matter in soils ranges from 100% in peats to less than 1% in desert soils. Kononova (1961) gives the following figures for the organic-matter content of various Russian soils as follows: podzolic soils, 3.5%; weakly podzolized soils, 5%; chernozems, 7.5–10%; dark-chestnut soils, 3.5%; sierozems, 1.5%; kraznozems (occur in Australia but not in the United States), 5.0%.

The humus in soils does not have the same composition in every region, but there is a regional pattern of humus content and composition in the various climatic zones. This regional pattern was first recognized by Tyturin (1949); it is caused by the length of growth period, by the water regime, and by regular periods of dessication (as in monsoonal climates). These variations cause changes in the intensity of the conversion of organic matter into humic substances. Different plants produce different amounts of raw organic matter, and in a different form. For example, deciduous trees produce large quantities of leaf litter; grass roots produce large quantities of organic matter distributed through the soil. The compositions of the humic substances active in weathering differ in the soils that have developed. And conversely, the type of organic matter has caused the formation of different soils in different regions. Kononova (1961) gives the amount of humic and fulvic acids in the organic matter of certain soils (Table 30).

9–4. Humus in Soil Formation

A soil without organic matter and humus is an abiotic soil and any mineralogical changes that take place on weathering are caused by inorganic chemical processes. The presence of organic matter, raw or converted to humus, is important in establishing and maintaining pedochemical weathering and in supplying carbon and nitrogen for the growth of microflora as well as macroflora. The complex organic matter that is produced by the stems, leaves, and other parts of plants is broken down into simple inorganic compounds such as carbon dioxide, water, and nitrates. The amount of change from raw organic matter to these compounds is measured by the C–N ratio which is routinely determined in soil laboratories as a characteristic of a soil. One of the most obvious functions of humus in weathering rock or soil is in causing aggregates and crumb structures to

TABLE 30
Composition of Humus in the 0- to 20-cm Layer of Soils in the USSR
(after Kononova, 1961, p. 233)

Soil type	Humic acid, %	Fulvic acid, %	Other constituents, %
Podzolic	20	47	33
Gray soils, weakly podzolic	25	50	25
Chernozem, leached	35	42	23
Chernozem, deep	40	39	21
Chernozem, normal	35	37	28
Chestnut, dark	34	35	31
Solonetz*	23	45	32
Sierozem†	21	41	38
Kraznozem‡	15	50	35

* In chestnut zone soil; solonetz is a soil developed in salty, arid areas with a friable surface underlain by hard, dark material.
† Sierozem is an arid-zone soil with a brownish surface grading down into a carbonate-accumulation layer.
‡ A loam or latosol consisting of deep, friable clay with little horizon development; develops on easily weatherable rocks under humid conditions; acid reaction.

form which permit water and air to circulate, and thereby assist plant growth. Roots cannot penetrate hard, compact clay. The comparison of a sticky clay before and after the addition of organic matter clearly demonstrates this point; long before scientific agriculture had developed, farmers were well aware of the value of organic matter to ameliorate or beneficiate soils, and it was from the use of farmyard manure that scientists discovered the important principles of ion exchange. By the combination of humus with clay minerals the humus is stabilized, and by cation exchange with plant root hairs nutrients are made available for the growth of plants. The importance of humus to soils and plants has been summarized by Broadbent (1962).

The role of humus in the two principal types of weathering situations, leaching and nonleaching, is shown diagrammatically in Fig. 26.

9–5. Complexing and Chelation (Chevulation)

Humic substances are high molecular weight compounds, as are protein, cellulose, lignin, starch, and others. They are the common substances of the organic matter that is added to weathering material by vegetation (and to a lesser extent by animals). Chemically they are ring compounds of either aromatic or aliphatic structure. Their chemistry and chemical reactions are complex. However, metallic cations are adsorbed by unsatis-

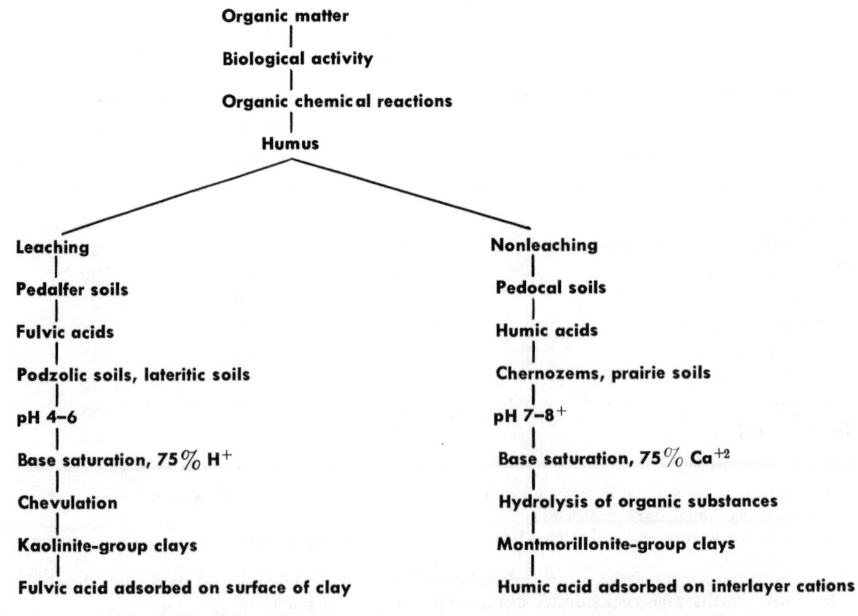

Fig. 26 The role of humus in weathering with and without leaching.

fied anions of the ring compounds. Chelation is one such method of adsorption. Chelation may be thought of as a coordination complex that can form with divalent and trivalent cations.

It has not yet been possible to isolate and identify all the organic compounds of humus. However, both humic acid and fulvic acid occur, as well as mixtures of the two. It was noted earlier that different climates cause a variation between humic and fulvic acid contents, and that these cause fundamental differences in the pedogenic weathering, i.e., in the soil formed. Excessive soil water (strong leaching in a moist climate) promotes the formation of humic acid with a small molecule, and hinders the growth of larger molecules. This occurs in podzolic soils in which the chief organic component of humus is found to be fulvic acid. There is a periodic dry season and the more complex humic acids are formed as in chernozems. The diagram above shows this relationship. The difference in water supply causes differences in the solubility of humic substances and in their reaction with electrolytes. It has been found (Kononova, 1961) that the greater the predominance of the aromatic ring in the molecules of humic acids, the less resistant they are toward electrolytes.

There are two groups of humic acids:

1. Brown humic acids—occur in brown coals, peats, and podzolic soils. These humic acids are very resistant to reaction with electrolytes.
2. Gray humic acids—occur in chernozems and rendzinas (black soils developed on hard limestone in poorly drained conditions). These humic acids have a high percentage of carbon and are readily coagulated with electrolytes.

Fulvic acids, approximating the first group above, are more stable than humic acids. They are very similar to humic acids but are soluble in water and in acid, whereas humic acids are not. Fulvic acids are the initial forms, or the decomposition products, of humic acids. Fulvic acids have many of the characteristics of crenic or apocrenic acids, and they behave like the simpler forms of humic acids.

Humic acids combine in various ways with the mineral constituents of soils such as calcium, the R_2O_3 group (iron and alumina), silica, and the clay minerals (in the latter they become structurally interlayered with the micaceous minerals).

Schatz et al. (1954) point out that most metallic cations in rocks and minerals readily enter into coordinate covalent complex formation with suitable organic chelators to form stable ring structures. They suggest that where acidity does not function in biological weathering, chemical degradation of rocks and minerals seems to be accomplished by chelation. Below pH 7 acids are important in biological weathering, but chelation is operative over a larger pH range.

Organic acids can form chelates with H^+, Mg^{+2}, Ca^{+2}, Sr^{+2}, Mn^{+2}, Co^{+2}, Cu^{+2}, Zn^{+2}, Fe^{+2}, Fe^{+3}, Al^{+3}, and V^{+4}. The chemical relations for solutions containing chelating agents are very complex (Martell, 1957; p. 16). It is probable that trace elements also form chelates.

Humus and other chelators in the soil help release and maintain trace elements in a form readily available to plants. Chelation is responsible for the ability of organic matter, specifically amino and hydroxy carboxylates to inhibit the precipitation of iron and aluminum as phosphates in soils.

A review of the investigations of soil organic matter up to 1954 has been given by Brenmer (1954).

The effects of chelation of iron and aluminum have been particularly studied in the development of the podzolic soil profile in which these elements are moved downward. Bloomfield has reported a series of experiments (1953 to 1954) with aqueous extracts of various plant leaves which were used to dissolve iron and aluminum under neutral and aerobic conditions. The ferric iron normally present in soils is reduced and changed to an organic complex that will slowly be reoxidized to the ferric condition. It is to be expected that this process will be at a maximum where organic matter accumulates, i.e., the cool, temperate climates, and at a minimum where organic matter decomposes readily and there is little leaf litter or mor horizon.

The pH of the extracts (5 g dry leaves in 200 ml of water) of the leaves investigated by Bloomfield (1953–54) are given in Table 31. These figures are comparable with those of leaf mold given by Jenny (1941, p. 226): norway pine, pH 4.5; white pine, pH 5.1; and maple–basswood, pH 6.5.

The term "cheluvation" was introduced by Swindale and Jackson (1956) to describe the process whereby minerals are decomposed by chelation and moved in solution to a lower part of the soil profile.

TABLE 31

Plant		pH
Scots pine needles		3.9
Agathis–Australis	young leaves	4.2
(Kauri)	old leaves	5.0
	bark	4.6
Dacrydium cupressium fronds (Rimu)		4.1
Larch needles		4.5
Aspen leaves		5.1
Ash leaves		5.6

Atkinson and Wright (1957), working with a 5×10^{-4} M solution of the disodium salt of ethylenediaminetetra-acetic acid (EDTA) at pH 4, formed a podzol-type profile in calcareous sand in the laboratory in about 17 months. The first action of the EDTA was to decompose the carbonates, thus releasing calcium and magnesium which moved downward in the "profile," probably both in the chelate and ionic forms. As the pH and the concentration of the alkaline earth metals in solution decreased at the top of the soil column, increased amounts of iron and aluminum were carried downward. With depth in the profile, higher pH values, and increasing competition from hydroxyl ions, it is probable that alkaline earths, iron, and aluminum are precipitated. The iron compound with EDTA is unstable at pH values between 7 and 8.

Organic acids that form chelates attack silicate minerals. Perkins and Purvis (1954) found that aluminum was chelated and potassium removed from orthoclase feldspar by EDTA treatment. Goldberg and Arrhenius (1958, p. 160) reported that EDTA attacked various minerals in the following increasing rate of dissolution: biotite, sphene, goethite, magnetite, rutile, hematite, lepidocrocite, nontronite, augite. Calcite and apatite were more strongly attacked than any of those listed. Tourmaline is also attacked by EDTA (Graham, 1957) and releases boron. It is probable that chelating can also partly account for the etching on the surfaces and ends of grains of highly resistant minerals such as staurolite, sillimanite, and garnet, and for the development of pits and cracks in zircon grains.

9–6. Organic Weathering

Organic weathering is an important process in the earliest stages of rock weathering of massive rocks in very cold climates. Jacks (1953) summarized the findings of a number of Russian investigators. Lichen and mosses are the first, and often the only, colonizers of rocks at high altitudes and latitudes. Acids from lichens decompose rock minerals which form a "synthetic clay–humus" that contains elements such as phosphorus, magnesium, calcium, potassium, sulfur, and iron that are necessary for the growth of higher plants. Kononova (1961; p. 167–171) gives further details of the weathering effect of these lowly plants. The process is a general one as analyses from the Arctic regions of Canada (Feustel and Dutilly, 1939) and from the higher parts of the Australian Alps show (Costin et al., 1952). In the latter locality, organic weathering is responsible for the development of alpine-humus soils at an elevation of about 5000 ft (the snow line). Analyses of the ash of the organic layer of these alpine-humus soils show that they have almost the same chemical composition as the parent rocks, which are gneissic granites.

Glazovskaya (1950) found that blue-green algae and diatoms were active weathering agents. The weathering products included amorphous silica and clay minerals. Glazovskaya (1958) also reported on the weathering processes in the Antarctic. These processes are almost exclusively inorganic, although a small population of bacteria exist and over 350 species of lichens (nearly all crustose, rock-inhabiting ones), constituting the major elements of the flora (Rudolph, 1966), are present. In addition there are about 200 species of freshwater algae, mostly inhabiting lakes, melt pools, and wet soil, and about 75 species of mosses in moist habitats as far south as 84°40′ S. This flora removes a small quantity of chemical elements, principally calcium, from the rocks on which it grows, and accumulates a small amount of organic matter in suitable microclimatic situations.

Blue-green algae are the first colonizers of calcareous rubble on coral atolls in the Pacific Ocean (Fosberg, 1953; p. 20), as they doubtless are elsewhere in suitable climates.

Chapter 10

TEMPERATURE IN WEATHERING

From the shape of the earth, its axis angle of 23½°, its tilting, and its rotation around the sun, it is evident that there should be seasonal variations in weathering produced at different temperatures. Although there are differences caused by topography and other factors in any particular area (Fig. 4, Section 2–2), the following major facts cause the weathering in different zones: high average temperatures in moist regions increase vegetation and chemical weathering; high average temperatures in arid regions decrease vegetation, decrease chemical weathering, and promote mechanical disintegration; low average temperatures in moist regions cause permafrost and soil freezing, and slow down soil-forming processes generally—water movement is prevented or only occurs in a thin layer above the permafrost in summer; low average temperatures in dry regions produce maximum physical disintegration, and chemical weathering is either nonexistent or very slight. As well as regional climatic differences there are regimes of soil climates which are in part described as microclimates. Each soil has a particular climate within which the soil flora and fauna live, and within which chemical changes take place. The soil climate is dependent on the overall climate of the region, but it is modified by the material of the soil itself, whether sandy or clayey, light colored or dark, and by the water regime. The soil climate is responsible for the chemical environment of the weathering processes. Farmers producing crops have always known from observation about variations in soil temperatures, some fields being "warm," others "cold." In recent years two important books describing soil climates and microclimates have appeared: *The Temperature Regime of Soils* by A. M. Shul'gin (translated from Russian by the Israel Program for Scientific Translations; 1965), and *The Climate near the Ground* by R. Geiger (Harvard University Press, 1965). There is now a wealth of information about the temperature in soils of the temperate zone, but not very much about tropical soil temperatures.

10–1. Weathering Temperatures

Geochemical weathering varies from physical disintegration at the poles to saprolite formation at the equator where warm, humid climates prevail. In pedochemical weathering, which is superimposed on geochemical weathering, soil climates develop within the soil-forming material. Temperatures within a soil profile are usually different from those of the air above the soil. The great soil groups are not only an expression of the steady state of weathering; they also express the fact that soil temperatures are within a given range over wide areas. There are many factors to be taken into consideration when considering the effect of temperature on weathering: vegetation, albedo, nature of the soil and its heating capacity, evaporation, and drainage being perhaps the major factors. Vegetation insulates the weathering surface from the sun's rays. Permafrost isolates the soils formed from the underlying weathering rock. According to Thornthwaite (1958), "of the energy that reaches the bare ground or a vegetative cover, some is reflected immediately back to the sky . . . less than 50% of the original total of insolation of the sun goes to heat the surface layers of soil or water, to heat the air in contact with the surface, and to be used in the evaporation of water." As the various layers of soil are heated, the chemical elements present in the soil are activated. The factors of heat and moisture balance are most important in a number of ways; for example, the soil's microflora and microfauna are dependent on the heat and moisture regime, and their metabolism conditions the breakdown of the organic matter in the soil.

Some figures for the annual insolation of a few localities are given in Table 32.

TABLE 32

Vegetation and Annual Insolation for a Number of Localities (after Thornthwaite, 1958)

		Annual insolation	
Locality	Vegetation	Hours	Maximum possible insolation, %
Yuma, Arizona	Desert	3900	89
Heluan, Egypt	Desert	3670	84
Kano, Nigeria	Steppe	3000	68
Kaduna, Nigeria	Savannah	2774	61
Yangambi, Congo	Forest	1860	42

The reflectivity, or albedo, of the land surface and its vegetation is very important in weathering. In dry areas a high albedo reduces the evaporation of soil moisture. The vegetation in dry areas, e.g., sagebrush and eucalyptus species, is nearly always light in color and has a high albedo in contrast to the vegetation in moister climates, e.g., oaks and other deciduous hardwoods.

There is a considerable difference in temperature between the air above a soil and that of the soil itself.

There is a time lag in the heating of a soil compared with that of the air, and in many places there is also a very marked seasonal difference. At Nairobi, Kenya, about 1 degree south of the equator (McCulloch, 1959), the surface of the soil varies during the year from 95°F to 105°F, whereas the soil at one foot below the surface varies between 65°F and 70°F. In Djakarta, Indonesia, 6 degrees south of the equator, the annual range of air temperature is 60°F to 90°F; the soil at 2 inches ranges from 70°F to 100°F, and that at 39 inches from 80°F, to 80.6°F (Mohr and van Baren, 1954).

The earth only receives about 40% of the energy radiated by the sun. In Table 33 Mohr and van Baren (1954) give the radiation received at different latitudes.

The difference between vegetative production in the tropics and that in the more temperate climates is not due to the maximum radiation intensity, but to the actual duration of that radiation. This period lasts for 12 months in the tropics as compared, for example, with only 4 months in northern Europe. Similarly, the duration of the radiation varies from northern Canada to the Gulf of Mexico. Agriculturally this duration is

TABLE 33

Radiant Energies in Calories Received at Different Latitudes (after Knoch, 1929)

	90° Pole	80°	70°	60°	50°	40°	30°	20°	10°	0° Equator
Annual average per day	366	378	417	500	600	694	773	830	867	880
Summer maximum per day	1103	1086	1038	1000	1015	1015	998	958	900	810
Winter maximum per day	0	0	0	50	180	326	477	627	745	863

expressed as the growing period, and is measured from the first frost-free day to the last.

The main source of heat entering the earth's crust is provided by solar radiation. The heat generated in the weathered material permits vegetation to grow and soils to develop. Heat is absorbed by the surface of the crust, transformed into thermic energy, and transmitted to the upper and then the lower part of the weathered material. At night this material radiates heat, the surface cools, and the cooling is transmitted downward. Thus, the surface, by absorbing and radiating heat, regulates the thermal energy and therefore the thermal regime. The thermal regime is expressed by the type of natural vegetation that grows. Under natural conditions, an equilibrium exists between the formation of organic matter by vegetation and its decomposition by microorganisms. This balance is determined primarily by climatic conditions. For example, a forest develops a floor of leaf litter (and its accompanying microfauna and microflora) that maintains its thickness. High average annual temperature in a moist climate encourages the rapid weathering of parent rocks and their soils. In general, the speed of chemical reactions approximately doubles for each rise of 10°C. All forms of chemical weathering are extremely rapid in warm, humid regions. Examples of this kind of weathering are common on the Hawaiian Islands. Vegetation provides the organic matter and organic acids that add to the effectiveness of inorganic weathering. A steady state of weathering, due to the interaction of a number of variables, prevails in each climatic zone. Pedochemical weathering is the manifestation of this steady state.

TABLE 34

Mean Monthly Air Temperatures and Mean Monthly Soil Temperatures at Cape Hallett (after Benes, 1960)

		Mean soil temperature, °F	
Month	Mean air temperature, °F	10 cm	50 cm
April	6.1	2.8	8.7
May	−11.1	−8.1	3.1
June	−9.2	−5.0	−1.2
July	−19.2	−2.4	0.3
August	−13.7	−21.2	−10.7
September	−14.9	−18.4	−12.1
October	−1.8	−6.4	−6.5
November	15.7	12.1	7.8
December	25.7	36.0	27.5

Probably the maximum chemical weathering conditions occur in the rain-forest of northern Suriname (Dutch Guiana). The air temperature in the forest is between 85° and 90°F, the rainfall is about 100 inches a year, and the humidity is 30–40%. Soil temperatures average about 25°C (78°F) (Sakamoto, 1960). The minimum conditions for chemical weathering occur in polar regions where the air temperature is below zero and the ground is perpetually frozen. The humidity is very low.

In spite of the prevailing low temperatures, heating of rock surfaces, moraines, and sand surfaces by the sun causes milder microclimates to develop. The ground surface, when warmed in this way, has a much higher temperature than the air.

Soil temperatures were measured at Cape Hallett, Antarctica, in1958 by Benes (1960). His figures are given in Table 34.

Chapter 11

TIME IN WEATHERING

The weathered surface of the lithosphere that we now see is the result of the long-continued interaction of Jenny's five variables on the exposed surfaces of the rocks that make up the earth. However, this definition does not emphasize the importance of topography and erosion that modify the process of soil-profile development.

11–1. Duration of Weathering Period

It is difficult to judge the age of the soil profile, used as the unit of soil maturity, because not all materials form soil at the same rate. On the one hand there are solid rocks of various composition (Chapter 1), and on the other hand there are the so-called "ready-made soils" such as volcanic ash, pumice, unconsolidated sands and clays, alluvium, and loess. In some regions, such as Alaska, upland silt has been deposited. This silt originated as frost-split grains of igneous and other rocks. Naturally all these unconsolidated materials are of a grain size that is more easily acted on by chemical solutions than are solid rocks, and therefore the time taken for profile development is shorter than it is for a profile to develop in a saprolite.

Because of geologic events the soils and saprolites that are now conspicuous were formed during the Tertiary period, generally from the Miocene or Pliocene. Unless weathering has been preserved, either by a continuance of previous conditions and no erosion, or by burial, all older weathering products have been removed by erosion. The material removed by previous erosion now forms present-day sediments, or is the raw material for such sediments. Some of these erosion products can be recognized as Pleistocene and later beds. Sands, alluvium, and colluvium remain as erosion products in many areas. Some bauxites and laterites have been recognized as preserved materials from a previous weathering (Patterson,

1967). Patterson (1967, p. 1) states "Bauxite deposits have formed chiefly by weathering aluminous rock; . . . most are residual accumulations for which most constituents of the parent rock other than alumina have been leached. Bauxite occurs in rocks ranging in age from Precambrian to Recent, and many deposits in the tropics are still forming."

Soils that developed prior to the last glacial epochs in the Northern Hemisphere have been largely removed by glacial action from the land surface. In North America, during maximum advance, practically all the continent was covered by ice north of a line extending from New York City westward along the Ohio and Missouri rivers, swinging northward nearly to the Canadian border in Montana and Idaho, and reaching the Pacific near Puget Sound. Isolated ice caps covered parts of the Rockies, the Sierra Nevada, and other high ranges. Soils that had developed in the northern areas were destroyed during the Pleistocene. The deep weathering that was present in the southern parts of the Appalachians was preserved as saprolites, with lateritic soils in some places, but no true laterite.

Time relationships of the various soils in the eastern United States were studied by Carter and Pendleton (1956). The ages of these soils can be inferred from the map in Fig. 27.

During the same period, in Australia, there was no glaciation except in the highest parts of the Southern Alps south of Canberra, the Federal capital. Most of Australia is a plateau, and laterite developed extensively under conditions that were undisturbed geologically for millions of years. Most of this laterite is considered to be Pliocene in age.

After the outpouring of the lower basaltic flows in Antrim, Northern Ireland, there was a long period during which undisturbed weathering of the surface of the basalt took place. The basalt was converted to a saprolite largely composed of kaolin (lithomarge) before its conversion to gibbsite. The resulting bauxites are a mixture of gibbsite and kaolinite, and were formed in the late Miocene or early Pliocene before they were covered by another flow of basalt.

The stage of development of a soil profile is an indication of the length of time that the same conditions of weathering have occurred in a region. Soil scientists recognize mature, immature, eroded, and other types of profiles in the field that are generally mappable units. The soil profile indicates the overall effect of pedochemical weathering; this is most readily recognized in terrains that contain only one type of rock, as in the Hawaiian Islands which are an example of progressive weathering of basaltic rocks. Because the island chain grew through volcanic activity from northwest to southeast, the rocks of the northwestern end of the chain have been weathered for a much longer time than those of the southeast. Thus gibbsite is common on Kauai, but red and brown soils are found on Hawaii.

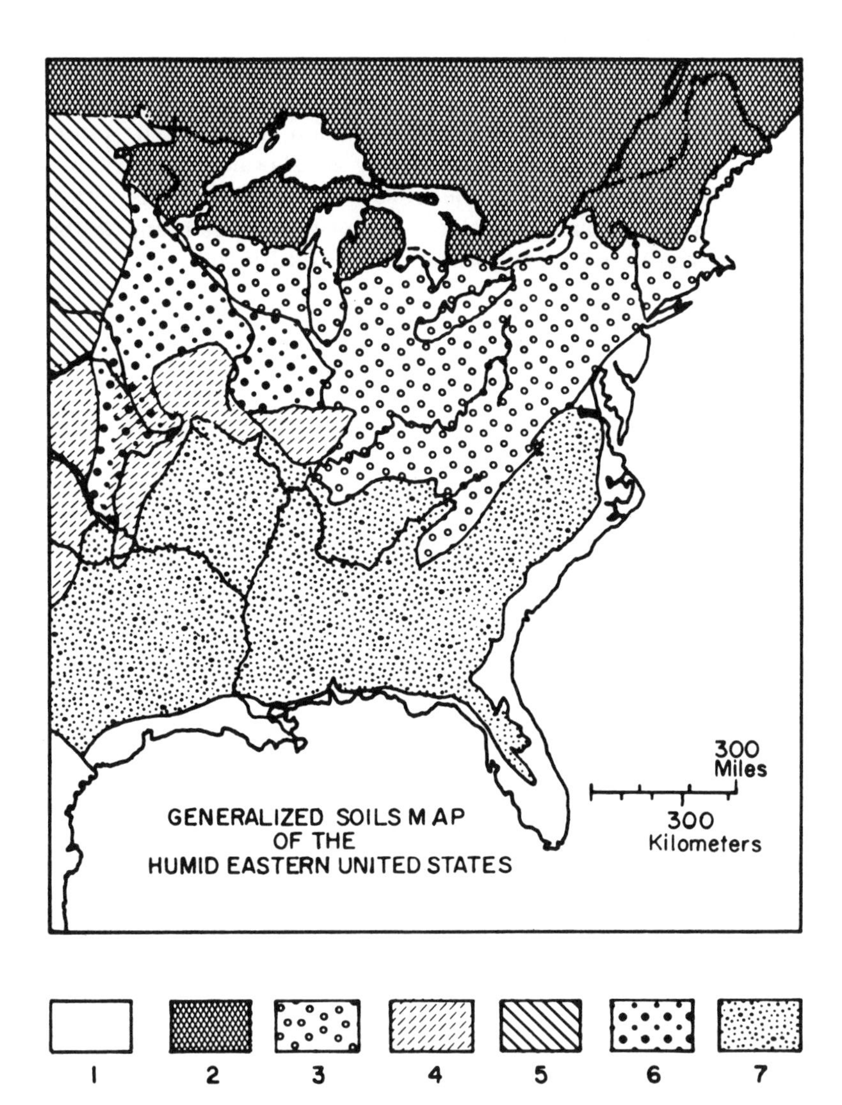

Fig. 27. Ages of soils in the eastern United States. (1) Groundwater podzols, wiesenboden, postdating interglacial high sea-stands. (2) Weak podzols, post-Valders. (3) Gray-brown podzolic soils, pre-Valders, post-Kansan, including Appalachian podzols. (4) Planosol, post-Nebraskan. (5) Chernozems, early Pleistocene. (6) Prairie soils, mostly pre-Valders. (7) Red-yellow podzolic soils, Pleistocene or earlier (after Carter and Pendleton, 1956).

Hough and Byers (1937) and Hough et al. (1941) of the U.S. Department of Agriculture published many chemical analyses of soils of different ages developed on Hawaiian basalt. Table 15 (Section 6–3) is an example of the chemical composition of soils of different ages developed on basalt. The weathering of volcanic material on Guam (Carroll and Hathaway, 1963) under conditions similar to those of the Hawaiian Islands provided an opportunity to compare the mineralogy of the Guam soils with those developed from basalt in other places. It can be seen from Fig. 28 that an age relation can be assigned to this group of basaltic soils.

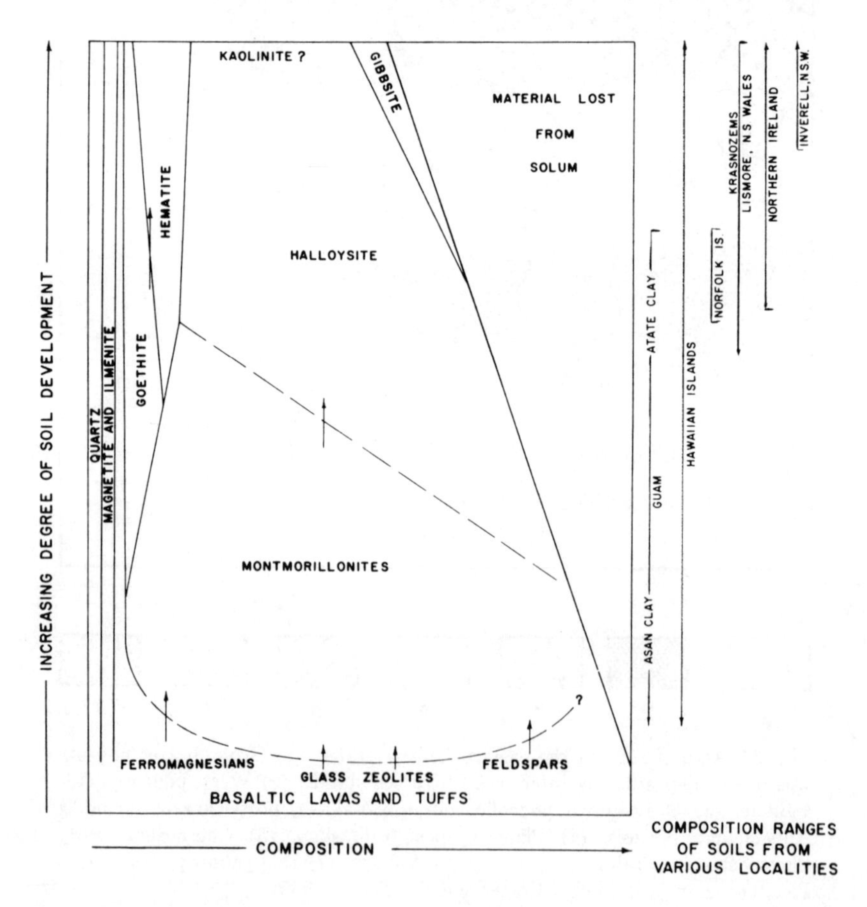

Fig. 28. Mineralogical relationships of soils developed on basalt in various localities (after Carroll and Hathaway, 1963).

11–2. Pedological Indication of Time

Some materials are porous and weather more easily than others. In solid rocks geochemical weathering appears to be very much the same in all humid climates. Other things being equal, saprolite formation will be faster in the tropics than in the temperate zones, but even so, it must be measured in geologic time. Porous materials are already halfway toward being soils and they do not form saprolites. Pedochemical weathering is expressed in the production of a soil profile (Chapter 4). If the profile reaches the maximum development for the climatic zone in which it occurs, the result is a mature soil. Soil scientists recognize maturity or immaturity in soils, but presuppose that the conditions at a soil site are suitable for the development of a mature profile before thinking in terms of age relationships. It is possible sometimes to differentiate between soils of various ages developed from flows of basalt, as in Hawaii. Hay (1960) found that a large fan of pyroclastic material (a ready-made soil) had weathered to a clayey soil in about 4000 years in St. Vincent, BWI. The weathered deposits were originally andesitic ash consisting of glass, plagioclase, hypersthene, augite, and olivine. Fine glassy ash altered to endellite (a type of kaolin), allophane, and hydrated ferric oxide. Some of the individual mineral grains have become etched in the weathering material. Hay considers that this 4000-year-old soil represents an early stage in the formation of the yellow earths of St. Vincent. The clayey soil formed from ash at an average rate of 1.5–2 ft per 1000 years. The glass decomposed at a rate of 15 cm (6 inches) per 1000 years.

In the upper Mississippi River basin (Leighton and Willman, 1950) windblown loess of several ages is extensively developed and has been correlated with the various glaciations of the Pleistocene. Most of the loess is calcareous, and when leached by rain the calcium carbonate is moved downward. Under original forest more $CaCO_3$ was removed than under grass. Smith (1942) found that the average loss of $CaCO_3$ in the Peorian loess (on which the Brady soil is commonly developed) during the period of deposition is equivalent to a layer of $CaCO_3$ particles about 30 inches thick, in which case the period of deposition was longer than the subsequent period of weathering. All the tills and loesses associated with the several glaciations were somewhat similar in this respect.

One of the most comprehensive studies of soil-profile development with age was made by Dickson and Crocker (1953) of soils developing under coniferous forest on the southeastern slopes of Mount Shasta, northern California, at a height of 3450–3850 ft. The parent material consists of a series of mudflow deposits more or less superimposed in a regular stratigraphic sequence. The mudflows consist of a mass of pulverized rock of hornblende–andesite composition. The ages of the four youngest flows were determined by the age of the oldest trees (growth rings) growing on the surface of the deposits. The age of these flows are: A, 27; B, 60; C, 205; and D, 566 years. The age of E was determined from morphological characteristics of the soil formed on it and buried by flow D 566 years ago. E is estimated to be 1200+ years old. In this area the average yearly rainfall is 46 inches, and the average snow cover is 102 inches. The average annual temperature is 49.6°F, ranging from 35.1°F in January to 65.9°F in July. The mineralogical and physical properties of all the mudflow deposits are similar, as are the slope and southeastern exposure of the sampling sites. During soil-profile development, various chemical processes occur (see Chapter 8); these are conditioned both by the inorganic matter of the soil and by the plants and animals that live there (the ecosystem). Although the results of the Mount Shasta soil study are complicated by a recycling of various chemical elements, it was found that the forest reached a steady state within the 1200+ year period. A forest floor was developed and this allowed organic carbon and nitrogen to become a part of the soils. A pH gradient was established by the deposition of the acid litter on the surface. It was found that the variability of the soils was small compared with the established ages of the andesitic mud. There was very little change in the minerals present in the material, although a very small amount of (montmorillonitic) clay did develop. The formation of clay minerals is extremely slow in this kind of material. No marked textural profiles were developed, and the authors concluded that geological, rather than historical, time would be required. Other soils of the area, presumably much older, show no pronounced textural profile. Organic matter became mixed with the mineral grains and changed the porosity from about 30% in the parent material to 60% in the surface of the older soils. Free iron oxide did not appear to accumulate until after the 205-year stage. The accumulation may be due to transport by chelation. This study shows quite clearly the interrelationship between "plant succession" and "soil" development, and emphasizes the fact that soil development is dependent on the pioneering and growth of plants.

11–3. Fossil Soils

It was mentioned in Section 11–2 that buried soils are found in many regions. Any soil which developed under conditions different from those of the present zonal soils of a region can be termed fossil. Such soils are recognizable in several ways:

1. In the field dissimilar soil profiles may be adjacent in an area in which all the soil profiles should conform to the zonal pattern. For example,

in some areas in northern New South Wales soils of basaltic parentage are found to be red, brown, or black; the red soils are fossil and have lateritic profiles, the black soils are the mature soils of the area and are similar to chernozems, and the brown soils are the developing soils that are too young to have acquired the profile of the black soils. A somewhat similar situation occurs in the Hawaiian Islands.

2. Complete soil profiles may be covered by later material. Examples of one kind of such soils are found in the north-central United States.
3. Soils developed during a previous geologic time may be left as erosion remnants in a landscape in which present soils are forming. Examples are the laterite soil profiles in southwestern Australia, and the red-earth residuals of Queensland.

Some of the most striking examples of fossil soils are those found in loess in the north-central United States. In 1951 Thorp *et al.* described and summarized much of our knowledge of buried soils in the central United States from the southern border of the last glacial drift (Mankato) to the Gulf of Mexico, and from the Appalachians to the Rocky Mountains. The buried soils of Nebraska, Iowa, Illinois, Kansas, and Missouri are the best developed of these soils that are associated with the four major Pleistocene glaciations, the Nebraskan, Kansan, Illinoian, and Wisconsin. The results of the advance and retreat of the ice cap during this period are complex, but Table 35 gives a generalized outline of geologic events during the Pleistocene period in Kansas (Zeller, 1968, p. 60)

Soil profiles have developed in the pre-Pleistocene loess and till that are sufficiently distinct to have received the names given in Table 35 and their recognition in sections through deposits has enabled geologists to establish stratigraphic relationships in the various loesses and tills. A road-cut with a fossil soil is shown in Plate 3B.

A soil profile developed on a weathered rock may be covered by deposits of sand, and some of the characteristics of the chemical composition of the original profile may be mixed with the surface sand. Profiles of basaltic tuff covered by sand have been described by Tiller (1957, 1958) from the Mt. Burr area of southeastern South Australia. Although the relationship between chemical elements supplied by the tuff and the weathering processes of the formation of a sandy soil on a basaltic-tuff profile are complex, the surface soils contain less SiO_2, but more Zn, Cu, Ni, Mo, and Mn than a sand. The buried-tuff profiles contain greater concentrations of these and other elements than do the overlying sands. The contents of the buried-tuff profiles are similar to those of basaltic profiles that are not buried. Other similar dual profiles occur where basalt profiles have been covered with Quaternary sands.

TABLE 35

Classification of Deposits in the Pleistocene Series of Kansas (after Zeller, 1968)

Time-stratigraphic units	Rock-stratigraphic units					
	Northeastern area		Southeastern area		Central and western area	
Recent	Eolian and fluvial deposits					
Wisconsin	Bignell formation	Fluvial deposits	Bignell formation	Fluvial deposits	Bignell formation	Fluvial deposits
	Brady soil					
	Peoria formation	Fluvial deposits	Peoria formation	Fluvial deposits	Peoria formation	Fluvial deposits
	Gilman Cañyon formation					
Sangamonian	Sangamon soil					
Illinoisan	Loveland formation	Fluvial deposits	Loveland formation	Fluvial deposits	Loveland formation	
					Crete formation	

<table>
<tr><td>Yarmouthian</td><td colspan="4">Yarmouth soil</td></tr>
<tr><td rowspan="5">Kansan</td><td>Loess</td><td>Fluvial deposits*</td><td rowspan="5">Fluvial deposits*</td><td rowspan="2">Sappa formation*</td></tr>
<tr><td colspan="2">Cedar Bluffs till</td></tr>
<tr><td colspan="2">Fluvial deposits</td><td rowspan="3">Grand Island formation</td></tr>
<tr><td colspan="2">Nickerson till</td></tr>
<tr><td colspan="2">Atchison formation†</td></tr>
<tr><td>Aftonian</td><td colspan="4">Afton soil</td></tr>
<tr><td rowspan="3">Nebraskan</td><td>Loess</td><td>Fluvial deposits</td><td rowspan="3">Fluvial deposits</td><td>Fullerton formation</td></tr>
<tr><td colspan="2">Iowa Point till</td><td rowspan="2">Holdrege formation</td></tr>
<tr><td colspan="2">David City formation</td></tr>
</table>

* Locally contains the Pearlette ash bed.

† The Atchison formation has been defined as proglacial outwash of early Kansan age. Similar deposits of sand are found between the Nickerson till and the Cedar Bluffs till.

In the most important loess area of the Netherlands, South Limberg (Edelman, 1950), two layers of loess are found, each having a soil profile of its own and separated by a vegetation horizon. Similar profiles are found in the loess areas of Belgium, northern France, and Germany. The original loess–loam profiles were slightly podzolized brown forest soils. The age of the buried profiles can be obtained from the stratigraphy of the loess, as it is in the United States.

Erosion remnants are particularly recognizable in arid or semiarid landscapes, and in some regions soils developed in a previous geologic time may remain *in situ* as complete profiles or as truncated profiles. The lateritic soil profiles that remain in southwestern Australia serve as an example. Probably the first geologist to recognize the true nature of the remnants of an extensive lateritic capping on saprolized country rock was Walther (1915), whose original drawing of such an erosion feature is illustrated in Fig. 11 (Section 5–4). These old profiles form mesas and buttes of a weathering and soil-forming period (thought to be Pliocene) that resulted in an indurated, concretionary ironstone capping that preserved the underlying geochemically weathered Precambrian gneiss from being entirely removed by erosion; the arid climate preserved the erosion remnants much better than would have a humid one with rivers. The capping had long been recognized as laterite, but it was due to Prescott (1931) that the laterite was recognized as the B horizon of a podzolic type of soil. Inherited features of these soils are the low pH (in a region of alkaline soils) and the almost complete removal of the major and minor chemical elements by leaching. The original surface of these soils was a sand which remains *in situ* in some instances, but has largely been removed and now forms sand-plains, or is added to the new soils forming in the country rock where the old profile has been completely removed. Large areas of the country are mosaics of sand, laterite, and new soils, and as such are difficult to develop agriculturally. Detailed descriptions of the removal of the original A horizon of these old laterite profiles have been made by Mulchay and Hingston (1961).

Chapter 12

TRACE ELEMENTS IN WEATHERING*

In addition to the principal elements that, with oxygen, account for over 99% of the composition of rocks—the silica, alumina, iron, and other oxides that are determined in the usual chemical analysis of a rock—there are minute quantities of other elements that are measured in parts per million (ppm) and known as trace or minor elements. These elements are interesting geochemically and are of considerable importance in plant and animal metabolism; some are necessary in life processes.

12–1. Amounts of Trace Elements in Rocks in the Lithosphere

The average amounts of many of the trace elements present in 99% of the rocks making up the lithosphere are given in Table 36. Some of the elements included in this table are in rather larger amounts than the majority. They have been included because of their role in plant and animal nutrition.

12–2. Association with Minerals

The primary sources of trace elements are magmas that on consolidation form igneous rocks. When magmas solidify the chemical elements, both major and minor (trace), crystallize as minerals, the trace elements in isomorphous replacement in the major elements. Some exceptions to isomorphous replacement do occur, one such being gold which may occur as metallic gold. The rock-forming minerals commonly act as hosts to the trace elements.

---

* Publication authorized by the Director, U.S. Geological Survey.

TABLE 36

Average Amounts in ppm of Some Trace Elements in Rocks of the Earth's Crust (after Krauskopf, 1967, p. 639, 640; courtesy McGraw-Hill Book Company)

Element	Atomic No.	Crust	Granite	Basalt	Shale	Ionic radius, Å
Li	3	20	30	10	60	0.68
Be	4	3	5	0.5	3	0.35
B	5	10	15	5	100	0.23
F	9	625	850	400	500	1.36
S	16	260	270	250	220	0.68
Sc	21	22	5	38	10	0.81
V	23	135	20	250	130	0.74^{+3}, 0.59^{+5}
Cr	24	100	4	200	100	0.63^{+3}, 0.52^{+6}
Co	27	25	1	48	20	0.72^{+2}, 0.63^{+3}
Ni	28	75	0.5	150	95	0.69
Cu	29	55	10	100	57	0.96^{+}, 0.72^{+2}
Zn	30	70	40	100	80	0.74
Ga	31	15	18	12	19	0.62
Ge	32	1.5	1.5	1.5	2	0.53
As	33	2	1.5	2	6.6	0.58^{+3}, 0.46^{+5}
Se	34	0.05	0.05	0.05	0.6	0.50^{+4}, 0.42^{+6}
Rb	37	90	150	30	140	1.47
Sr	38	375	285	465	450	1.12
Y	39	33	40	25	30	0.92
Zr	40	165	180	150	200	0.79
Nb	41	20	20	20	20	0.69
Mo	42	1.5	2	1	2	0.70^{+4}, 0.62^{+6}
Ag	47	0.07	0.04	0.1	0.1	1.26
Cd	48	0.2	0.2	0.2	0.3	0.97
Pb	82	13	20	5	20	1.20^{+2}, 0.84^{+4}

In isomorphic replacement the following conditions must be fulfilled:

1. The ionic radius of the replacing ion must be about the same size as that of the one replaced (there is a tolerance of about 10%); for example, Fe^{+2} with an ionic radius of 0.83 Å can be replaced by Co^{+2}, having an ionic radius of 0.82 Å.
2. Replacement takes place more readily if the charges of the two ions are similar, as they are in the example given.
3. The coordination number, i.e., the number of ions bonded to the replacing ion, must be similar. For example, Ba^{+2} with coordination 8 cannot replace Fe^{+2} with coordination 6, even though the size difference makes this replacement impossible.

An additional factor, implied by the coordination number, is that the resulting structure must preserve electrical neutrality.

The crystal structures of tremolite (a pyroxene) and olivine serve as examples of isomorphic replacement by trace elements. The pyroxene structure consists of single chains of SiO_3 groups. The formula for tremolite is $(OH)_2Ca_2Mg_5Si_8O_{12}$. In this structure there are seven cations which can be replaced by others of a suitable size; two of these are Ca, ionic radius 1.06 Å, and five are Mg, ionic radius 0.78 Å. Cations suitable to replace Mg are Fe, Co, Ni, Zn, and Cu, all of which are divalent. These elements have been recorded as traces in many analyses of pyroxenes. Olivine has a much simpler structure; it consists of separate SiO_4 groups packed closely together with Mg ions between 6 oxygen atoms. Mg is readily replaced by Fe, and analyses show that Ni and Cr also replace Mg.

12–3. Distribution in Granitic and Basaltic Rocks

As already noted there are two principal groups of igneous rocks: (1) granitic, which are siallic, high in silica, and contain quartz; and (2) basaltic, which are basic and mafic, much lower in silica than granitic rocks, and only have silica combined in silicate minerals. The minor elements associated with granitic rocks are Li, Be, B, F, S, Ga, Rb, Y, Zr, Mo, and Pb; those associated with basaltic rocks are Sc, V, Cr, Co, Ni, Cu, Zn, Sr, and Ag. Shale represents the sedimentary rocks that have formed as a result of erosion of igneous rocks. The average shale has been found to be enriched in Li, B, Ni, Cu, Zn, Ga, As, Se, Rb, Sr, Zr, and Pb when compared to the average trace element content of the earth's crust.

The granitic and basaltic rock-forming minerals with their associated minor elements are:

Granitic rock minerals	*Basaltic rock minerals*
Quartz	
Feldspars: orthoclase, microcline, albite to oligoclase, microperthite (Rb, Ba, Sr, Cu, Ga, Mn, Pb)	Feldspars: labradorite, bytownite, anorthite (Sr, Cu, Ga, Mn)
Micas: muscovite (F, Rb, Sr, Ga, V); biotite (Rb, Ba, Ni, Co, Sc, Li, Mn, V, Sn)	
Ferromagnesian minerals (minor): hornblende (Ni, Co, Mn, Sc, Li, V, Zn, Cu, Ga); augite (Ni, Co, Mn, Sc, Li, V, Zn, Pb, Cu, Ga)	Ferromagnesian minerals (major): augite (as for granite type); olivine (Ni, Co, Mn, Li, Zn, Cu, Mo)

12–4. Chemical Characteristics

The chemical characteristics of elements indicate in a general way how each will behave during inorganic chemical weathering. In the presence of organic matter and its alteration products the chemical elements have a somewhat different reaction pattern (Section 9–5). The elements can be grouped according to their ionic potential (Z/r, i.e., ionic charge or valence/ionic radius) as in Fig. 20 (Section 8–1).

Group 1: elements with Z/r up to 3.0; cations which remain in ionic solution during weathering and transportation, and may form basic oxides.

Group 2: elements with Z/r of 3.0–9.5; amphoteric and are precipitated by hydrolysis, their ions are then associated with hydroxyl ions from aqueous solutions. The line at Z/r of 9.5 represents the upper limit of elements concentrated in bauxite.

Group 3: contains only Si and Mo with ionic potentials between 9.5 and 12. The position of Si is anomalous; it is only very slightly soluble as the oxide, silica (quartz), but nearly 25 times more soluble as amorphous silica (opaline silica). When silica is in solution it ionizes as do the Group 1 elements (see also Section 8–3). Molybdenum exists in several valence states and therefore is also present with the Group 2 elements.

The ionic potentials of the elements show that Group 1 elements will go into solution and therefore will be largely removed from the weathering environment, and that Group 2 elements and Si will accumulate in the products of weathering. Goldschmidt (1937) recognized the importance and usefulness of summarizing chemical behavior according to ionic potential; Fig. 20 is a modification of Goldschmidt's original diagram. The examination of the minor element content of Arkansas bauxite illustrates the relationship of Groups 1 and 2 in hydrolyzate deposits (Gordon et al., 1958; p. 97).

The minor elements in Arkansas bauxite that were concentrated in comparison with those in the parent nepheline syenite (as measured by the Al concentration) are as follows:

$<$Al concentration: Zr, Ti, Sc, V, Be, Mn, Y, Pb
$>$Al concentration: Cr, Cu, Ga, Nb, Mo

Elements that were found to be less concentrated in the bauxite than in the nepheline syenite are Sr, La, Ba, Ca, and Mg.

12–5. Behavior of Trace Elements in Weathering

The stability series for the rock-forming minerals obtainable from the weathering potential index (WPI) (Section 4–4) indicates the way in which these minerals are expected to alter by weathering and thereby release trace elements to the weathered material, both saprolites and soils. The series can be written thus:

orthoclase > albite > oligoclase > andesine > labradorite > muscovite > biotite > hornblende > augite > olivine

It has been calculated that the WPI of granite is 7 and that of basalt 20, that is, basalt weathers almost 3 times as fast as granite under the same conditions.

When the host mineral is altered by weathering the trace element is released from its structural position. Its fate depends on the chemical reactions that take place with water or with a weak solution containing various ions. The behavior of the trace-element cation depends on its ionic radius and charge as shown in Fig. 20. The cation may go into solution, be insoluble, or partially soluble and concentrate in definite parts of the weathered material. The initial weathering is geochemical and inorganic, but as weathering proceeds and pedogenic weathering forms soils, the weathering environment becomes partially or wholly organic under the influence of vegetation. Trace elements may become associated with weathering products, clay minerals, or iron oxide nodules.

In geochemical weathering a trace element may remain in its host mineral (whether or not it does will depend on the stability of the host in the weathering environment), be removed wholly or in part in solution (if the drainage is good), or become associated with clay minerals formed during weathering. The completeness of the weathering depends on the type of rock, the climate under which it weathers (Tables 2 and 3), and on the effects of the five variables in weathering (Chapter 2). In other words, it depends on the steady-state chemical environment under which weathering takes place in geological time.

One important result is that although geochemical weathering may be complete in that a saprolite results, the soil formed from it may contain some unaltered mineral grains. The intensity of weathering, governed by time and erosion, will give the various amounts of sand, silt, and clay in the resulting soil (Fig. 19). Where chemical weathering is at a minimum in regions near the poles, physical weathering is at a maximum. The coarser in grain size the weathering product, the greater its resemblance to a crushed igneous rock, and the less dispersed the trace elements. This also applies to desert weathering in which drainage seldom, if ever, occurs. Investigation of the trace-element content of sand (2–0.02 mm), silt (0.02–0.002 mm), and clay (less than 0.002 mm) has shown that the sand and silt fractions of a soil show a greater quantity of trace elements than the clay. This applies particularly to soils developed from basic rocks in which there is no quartz. A large part of the coarser material of weathered granitic rocks consists of quartz sand. These two types of rock contain entirely different types of trace elements, those of basaltic rock being largely of elements necessary in plant and animal nutrition. The trace-element content of granitic rocks is geochemically interesting, but that associated with basaltic rocks is essential for nutrition. The coarser fractions of both granitic and basaltic rocks act as reserves or banks of trace elements which will become available as weathering continues in geologic time. The maximum leaching conditions of the warm, humid tropics will remove vastly more of both major and minor elements than the weathering in the temperate zones; the reserve of trace elements in the temperate-zone soils is greater than in the tropical soils. Table 4 gives an example of the kinds and amounts of residual minerals found in soils developed from limestone in Pennsylvania. In those soils the only minerals that have a low stability are feldspars and hornblende. Chlorite is probably derived from hornblende; muscovite, under normal weathering conditions is rather stable and probably only alters to an illite-type mica. These soils, therefore, probably obtain their trace elements from the reserve of hornblende, chlorite, and clay. In an investigation of residual minerals in soils (Carroll, 1945) it was found that soils developed on gneissic granite contained only 200–300 lb of amphibole and epidote per acre of the fine sand, whereas soils developed on dolerite contained over 1000 lb of epidote, 950 lb of amphibole, and nearly 9000 lb of augite in the fine sand of an acre of soil. The soils developed on gneissic granite were found to be cobalt deficient, whereas the soils developed on dolerite were nutritionally sound and supplied Co, which was the limiting factor in the growth of cattle feeding on the herbage of the soils derived from granitic rocks. The augite apparently acted as host for the cobalt.

12–6. Petrographic and Biogeochemical Provinces

A petrographic province is a large region which contains igneous rocks of approximately the same age that have originated in the same magma. In such a province all the igneous rocks have chemical and mineralogical characteristics by which they can be distinguished from apparently similar rocks from other petrographic provinces. The basaltic lavas of the Hawaiian Islands, the Columbia River Plateau, the Deccan in India, and the Antrim basalts in Northern Ireland, are examples of basaltic petrographic provinces. There are also numerous petrographic provinces of granitic and associated acid rocks, examples being the Boulder batholith, the New England granites, granites of the Mourne Mountains in Northern Ireland, and those from near Dublin in Ireland. The lavas of Mt. Vesuvius are distinctive in containing leucite.

Geologic mapping identifies the different kinds of rocks exposed on the earth's surface, and the distribution found forms a pattern of chemical, mineralogic, and physical composition; this pattern and its interpretation identify a petrographic province for major and minor elements. Thus a granitic terrain is able to supply those chemical elements and minerals that are associated with granites (Section 12–3), and a basalt; those associated with basaltic rocks. In addition to petrographic provinces that are recognizably associated with igneous rocks, provinces also occur in both sedimentary and metamorphic rocks. As sedimentary rocks are largely derived from the erosion, transportation, and deposition of the weathering products of igneous and metamorphic rocks, they contain the winnowed products of the distributive province from which they have originated, modified by the conditions of sedimentation. The major features of petrographic provinces were early recognized by land-use practices, but it is only recently that detailed chemical investigations of sedimentary provinces have been made. One such is the investigation of Mo in lower Liassic sediments, the parent material of soils in southwestern England; Mo toxicity is evidenced in cattle pastured on the herbage growing on these soils. The sediments nearer the source contain more Mo and are more toxic to cattle than those of similar age but of different source (Le Riche, 1959). The minor elements in the Triassic Jura Basin in France were found to be distributed as a function of distance from the Jura coastline and of the mineralogical composition of the clay and silt fractions in the sediments (Lucas and Ataman, 1968).

Both major and minor elements in petrographic provinces of regionally metamorphosed rocks, such as gneisses of Precambrian terrains, are inherited from the materials which were metamorphosed. Thus if the parent material of a gneiss was a resistate sediment, such as an impure sandstone, one would expect most chemical elements to have been removed by erosion prior to metamorphism. Petrographic provinces of metamorphic rocks are the result of numerous variable features.

Biogeochemical provinces exist because the chemical and mineralogical characteristics of petrographic provinces persist to a greater or lesser degree in the saprolites and soils developed according to the climatic zone in which the weathering occurred, subject to topography. The realization that biogeochemical provinces exist is due to A. P. Vinogradov, a Russian geochemist whose work followed that of V. I. Vernadski who gave the name biogeochemical processes to geochemical processes in which organisms participate (Vinogradov, 1964; p. 317). Biogeochemical provinces are very important for two reasons: (1) they are fundamental to animal and, partly, to plant nutrition, and (2) they are useful in geochemical prospecting for minerals that are of economic use. The investigation of biogeochemical provinces has been carried out mainly by Russian geochemists, and unfortunately most of the results of investigations are published in the Russian language. However, an English summary of the Russian investigations has recently been published (Malyuga, 1964). Vinogradov (1959, 1964) has applied the name biogeochemical provinces as follows:

> "We apply the name biogeochemical provinces to regions on the earth which differ from adjacent regions with respect to the content of chemical elements, and in which as a result experience a different biological reaction on the part of local flora and fauna. In extreme cases, as a result of a sharp deficiency or excess in the content of any chemical element (or elements) the plants and animals will experience biogeochemical endemias within particular biogeochemical provinces. By regions of the earth we mean rocks, soils, and water basins, defined and limited in time and space . . . populated by organisms. With respect to the content of one or more chemical elements in a particular region, that is, in rocks, soil, water, and the organisms of this region, we have in mind either the usual normal level or an inadequate or excess content of a particular chemical element of several elements (or their compounds). It should be noted that it is of importance to know not only the absolute content of elements in a particular region, in its rocks and soils, but also the content of any particular form for this element, for example, Fe^{+2} or Fe^{+3}; Mn^{+2} or Mn^{+4}; Se^{+2} or Se^{+4}. Thus, we often encounter not only an absolute shortage or excess, but also a relative deficiency or excess, depending on the degree of accessibility of any particular chemical element to the particular types of flora and fauna. For instance, Co and many other chemical elements in areas where there is an alkaline soil reaction, for example as a result of calcification, are quite inaccessible to plants. On the other hand, under these conditions Mo is very accessible to plants as a result of the formation of easily soluble molybdates, while in an acidic medium Mo is scarcely accessible to plants due to its leaching into compounds that are not easily soluble. The biological reaction of flora and fauna of a particular region, arising under the influence of an excess or deficiency of any particular chemical element (or elements), is the most important and basic criterion for the definition of a biogeochemical province."

Biogeochemical provinces are the natural state of the weathered surface of the earth, whether these provinces result from igneous, metamorphic, or sedimentary rocks. Vinogradov (1964; p. 334–335) describes sedimentary petrographic provinces as palaeogeochemical. The earth's crust has approximately 4×10^{12} km^3 of sedimentary rocks, or 5% of the crust $\times$ a depth of 10 km. The distribution of chemical elements in these rocks is governed by the amount and type of denudation of the land surface. The endemic flora and fauna originate from the living conditions of the petrographic provinces, both chemical and climatic; they identify these provinces as biogeochemical provinces.

Biogeochemical provinces, although they exist for most petrographic provinces, may not be easily recognized because the kind and amount of trace elements are not critical in plant, animal, and human nutrition. However, in many regions biogeochemical provinces are clearly defined. An example of a small biogeochemical province is a coral atoll in which there is an abundance of calcium carbonate, very little iron or alumina, and trace elements largely derived from corals and other marine organisms. The trace elements present in some of the atolls of the Marshall Islands, central Pacific Ocean, were found to be Cu, Mo, Fe, Cr, Ba, and B; in a few samples from these atolls very minute amounts of V, Ti, and Co were also identified in the spectrographic analyses (Fosberg and Carroll, 1965). Vinogradov (1964) cites a number of readily recognizable biogeochemical provinces: near Lake Victoria in Africa, the volcanic regions in Iceland, and isolated valleys in mountainous regions where only the exposed rocks provide the necessary chemical elements for plants and animals. Biogeochemical provinces are often strikingly shown by the flora that has developed, such as chalk flora and serpentine flora. Other biogeochemical provinces occur in areas that receive continuous annual increments of sodium chloride and have poor drainage, with deposits of volcanic ash which add needed chemicals in Indonesia but are deleterious in New Zealand, with the Pierre shales that add selenium in the northern part of the western Great Plains in the United States, and many others in most countries.

Biogeochemical methods of prospecting depend on the association of certain plants with certain chemical elements. In the United States identification of the presence of an ore body through plants is described as geobotanical prospecting (Cannon, 1960). Plants are used as indicators of certain chemical elements that are present in the weathered material above an ore body or as indicators of an enrichment of a chemical element in a sedimentary rock as, for example, uranium associated with the rocks of the Colorado Plateau. Plants may accumulate chemical elements or the presence of certain species of plants may indicate a weathered ore body. The soil and weathered rock of an ore deposit spreads out from the deposit as an aureole which can be mapped from the chemical analyses of the elements present, or by the association of certain plants with the soil enriched in these chemical elements. The extensive root systems of trees bring chemical elements to the sap whence they are distributed to the leaves, twigs, stems, wood, etc. For example, all plants contain small amounts of uranium (0.2–1.0 ppm), but the uranium content of plants growing in the residuum of rock enriched in uranium is much greater

(1.0–100 ppm) (Cannon and Kleinhampl, 1956; p. 682). Malyuga (1964) gives numerous descriptions of biogeochemical prospecting of ore bodies and states that deposits of Cr, Zn, Cu, Fe, W, Sn, Mo, Ni, Pb, and Be have been found by this method. He cites (1964, p. 3) the indicator plants or the humus layer of the soil through which the ore body was located.

12–7. Relationship of Trace Elements to Plant and Animal Nutrition

One of the most interesting studies of geochemistry is the very close relationship between trace elements and plant and animal nutrition. This relationship has been known for a very considerable time, especially for such elements as iron and iodine; animal malnutrition had been recognized by farmers who realized that certain fields would produce good animal and crop growth whereas others would not. However, the use of land for agriculture in Australia and New Zealand concentrated interest on many nutritional deficiencies that were not always so apparent in lands that had been settled for centuries. The study of animal nutrition in particular is a 20th-century achievement. It has been accomplished by the development of chemical analytical instrumentation that enables very small amounts of elements to be determined accurately, and by the use of radioisotopes so that the distribution of elements in the animal body can be determined. These new techniques have been used to obtain results from rigorous and sophisticated investigations in many countries (Underwood, 1962).

The rocks of the lithosphere supply all the micronutrients required by plants and animals with the exception of oxygen and elements supplied from the sea. Plants grow in soil and absorb the chemical elements present in that soil; they are seldom selective in the cations that they absorb. If the parent material contains no cations of a certain chemical element then the plants growing on the soil formed from that parent material will likewise contain none of these elements. The trace elements present in rocks when weathering and soil formation has occurred are absorbed by plants through exchange reactions between the colloidal part of the soil and root hairs. Some of the trace elements absorbed by plants are necessary for growth, but plants will absorb any trace elements that are present in the soil. Some kinds of plants are able to accumulate certain elements; for example, silicon is accumulated in grasses where it gives rigidity to stems. The very large number of trace elements that have been determined in soils can be arranged in three large groups:

1. Important in plant, animal, and human nutrition—B, Co, Cu, Mn, Mo, Zn, Se, I, Na, Cl, F, S, and V. Iron, although not a trace element, is also necessary.

2. Important in geochemical prospecting—Au, Ag, Co, Cu, Ni, Pb, W, Zn, Sn, and U.
3. Present, but of unknown importance—Ga, Li, Rb, Sc, Cd, Cs, Ge, Hg, Nb, Ta, Sb, Zr, and others of rarer occurrence.

The trace elements found are measured in parts per million (ppm) except for some of the essential elements that are present in rocks in larger amounts, such as Fe, Ca, Mg, and others. The approximate quantities of trace elements that have been recorded in soils are shown in Fig. 29.

That environmental factors are important in human nutrition is illustrated by the incidence of goiter in certain areas of the world (Underwood, 1962; p. 221). It is becoming increasingly evident that environment factors (micronutrients) are important in a number of human malnutrition disorders, some amounting to diseases. At a meeting of the Geochemical Society (U.S.) in 1963 (Cannon and Davidson, 1967) it was stressed that cooperation between geochemists, soil scientists, and plant, animal, and human nutritionists was needed to study the effects of environment on nutrition, and particularly the distribution of trace elements in geological materials and their dispersion on weathering. Cooperation is continuing to progress in these fields since that initial meeting.

Plant Nutrition

Plants absorb chemical elements from the soil in which they grow. Many elements are associated with the parent rocks of soils, and plants

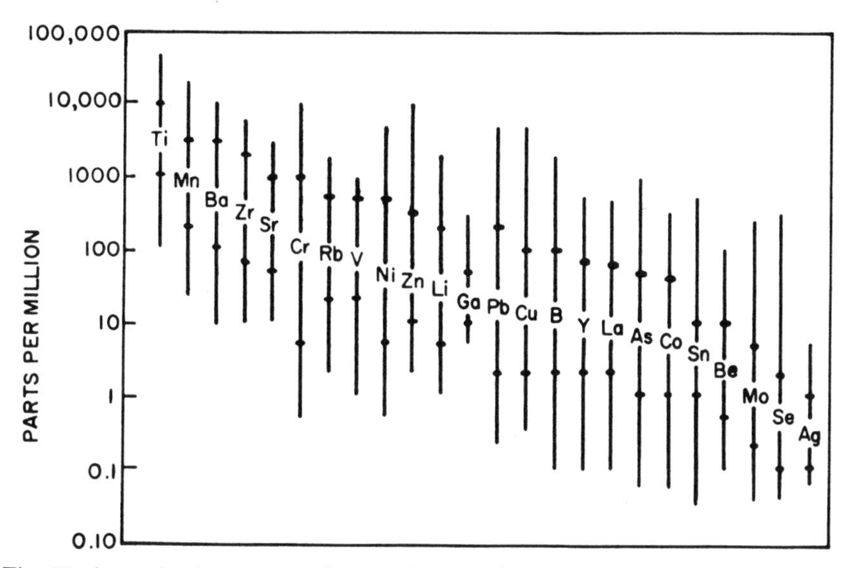

Fig. 29. Approximate amounts of trace elements that have been determined in soils (after Swaine, 1955).

have little discrimination in the elements they absorb. Some sixteen elements are necessary for the growth of plants (Dean, 1957)—C, H, O, N, P, S, K, Ca, Mg, Mn, Zn, Cu, Mo, B, Fe, and Cl. The macronutrients, C, H, and O, are obtained from water and carbon dioxide, and the remainder from the soil. The micronutrients are Fe, Mn, Zn, Ca, Mg, Mo, B, and Cl. Green plants also contain Na, I, and Co, which are essential to animals but not, with the exception of Co, necessary for plants. As plants draw nourishment from soils they may accumulate chemical elements to such an extent that ingestion of the plants by animals produces a toxic effect. An example of this is the accumulation of Se by certain plants in dry climates.

The nutrient supply of plants comes from the soil water which has reacted chemically with mineral grains in the solid part of the soil. In the form of cations in the soil solution, the chemical elements are attracted to the negatively charged cells of the root hairs where they are absorbed into the plant's sap by cation exchange. Many of the cations originally combined in minerals in the parent rocks become associated with clay minerals when the original minerals are altered by weathering. The clay minerals, through cation exchange, are able to release cations to the soil solution and, therefore, to the root hairs very readily: hence the value that has always been placed on the clay content of soils.

Iron and manganese are generally present in comparatively large amounts in soils, but both may be in very small quantities in soils formed from sandstones.

Iron. Iron, although present in soils in abundance, may not be in a form available to plants. Its availability is insufficient in the presence of an accumulation of Cu, Mn, Zn, and Ni, and lime-bearing soils generally contain too little. In the United States Fe deficiency is common in intermountain soils and in those of the southwest, that is, in unleached soils. One of the functions of Fe is as a catalyst in the production of chlorophyll, the green coloring pigment of leaves. Fe-deficient plants have yellow leaves and are chloritic. The active Fe in plant metabolism affects the Fe–porphyrin protein complex which acts as an oxygen carrier. Fe becomes active biologically only when it becomes part of these organic compounds, and factors that limit the supply of Fe to the plant limit the life of the plant. Fe may be leached out of soils by the chelating action of organic acids such as occur in peats (Section 9–5), peat water being generally brown in color, or when rain is abundant in a tropical region. In the latter condition bleaching of the saprolite below a soil generally occurs (Fig. 11, Section 5–4). The relationships of Fe in soils are complex and often oxidation and reduction are present. Fe oxidizes readily and in soils is normally in the ferric form. As ferric oxide it adheres to clay and mineral particles, but it is reduced to ferrous oxide or hydroxide in gley soils and in rice paddies. Plants vary in their resistance to chlorisis which may be corrected by application of iron salts in an available form such as iron chelate. (Holmes and Brown, 1957).

Manganese. Plants require minute amounts of manganese to grow properly, but Mn-deficient soils are found in many areas in the United States and elsewhere. Mn acts with Fe in the synthesis of chlorophyll, and Mn deficiency causes a chlorisis that has the appearance of Fe chlorosis. The factors that influence the availability of Mn to plants in the soil are tied up with its chemistry and with biological factors. Soils with pH above 6 favor the oxidation of Mn^{+2}. The oxidation–reduction conditions in acid soils favor the reduction of Mn^{+4} to Mn^{+2}, an available form. Mn can be leached from an acid soil but not from an alkaline. The exchangeable and water soluble Mn (Mn^{+2}) is the only form available to plants. The manganese cycle in the soil is (Sherman, 1957)

$$Mn^{+2} \rightarrow \text{colloidal hydrated } MnO \cdot MnO_2 \rightarrow \text{inert } MnO_2$$

Copper. A deficiency of Cu reduces plant growth, but an excess is toxic. The almost complete failure of crops to grow on the peat soils of the Florida Everglades was caused by Cu deficiency and was cured by the application of Cu as a fertilizer. The functions of Cu in plant nutrition are many and varied; Cu is concerned with the rootlets of plants and has an important function in the metabolism of roots, it appears to be related to the utilization of ammonium nitrogen by plants and to that of proteins in plant growth, and it is important in oxidation–reduction reactions. Cu functions as an enzyme activator and as an integral part of the enzyme molecules. It is difficult to ascertain the amount of Cu in a soil that is available to plants (Reuther, 1957).

Zinc. Zinc is necessary to plants, but there are many soils that contain insufficient available Zn, such as peat soils and calcareous soils. The Columbia River Basin in Washington and the Sacramento Valley in California contain soils that are naturally low in Zn. Fixation of Zn in an unavailable form commonly occurs. Zn is concentrated in the surface soil and the amount declines with depth. Plants remove Zn from the subsoil, transport it to the leaves, and concentration occurs in the leaf litter. The availability of Zn is at a minimum for pH 5–7, but is more available in acid soils. Zn is adsorbed on clay and may be fixed in clay minerals, as the Mg and Zn ions are approximately the same size. Zn also interacts with phosphate in soils forming an insoluble Zn–phosphate complex, and it is also commonly adsorbed on calcite and dolomite. Zn deficiency in soils in California has been linked to a high organic-matter content. Zn deficiency in plants may lead to cell enlargement, reduction in the number of chloroplasts, and the absence of starch grains, among other symptoms. A deficiency of Zn may affect the uptake of other micronutrient cations by plants. Zn is necessary in several enzyme systems that regulate the plant's metabolism; for example, it regulates the equilibrium between carbon dioxide, water, and carbonic acid, and it is associated with the water relationships in plants. Zn deficiency does not develop so rapidly in mild sunlight as it does in bright sunlight in areas where Zn has a low availability in soils. Plants differ in their efficiency in extracting Zn from soils, and consequently many trees are liable to show symptoms of Zn deficiency (Seatz and Jurinak, 1957).

Boron. Many soils in the United States are boron deficient, 41 states having reported boron deficiency for about 90 crops. Boron was first used as a fertilizer about 400 years ago when borax was shipped from central Asia to Europe, but it was not thought to be necessary until 1915. The first proof of boron deficiency came from the Rothamsted Research Institution in England in 1923. Most of the boron in soils is present in the mineral tourmaline, a very complex aluminoborosilicate, that is a minor mineral in pegmatites and sometimes in granites. This mineral is one of the least easily altered by weathering and, with zircon, may be the only detrital "heavy" remaining in geologically old, reworked sands and sandstone. The boron content ranges between 20 and 500 lb in the ploughed layer of an acre, but only about 5% of the total is available for plant use. Available boron occurs in inorganic and organic forms. Microorganisms and plants use inorganic boron for growth and change it into the organic form. Graham (1957) showed that boron forms a chelate with EDTA and, in the weathering of tourmaline, chelation is probably one way that boron is removed from its host mineral. Boron is lost from the soil by removal by crops. Plants are variable in their boron requirements; thus 1 ton of alfalfa hay contains 1 oz of boron, whereas a ton of sugar beets contains 2.5 oz. Leaching losses of boron are considerable, particularly in acid soils of humid regions. Light soils (with a porosity that allows for water circulation) remove more boron than do heavier clay soils. Fixation of boron is closely related to clay content and to soil pH, and increases as the soils become more alkaline. Silt-sized minerals and organic matter also fix boron. The most soluble boron in soils is the organic form, but when a soil is dried this boron is increasingly fixed and becomes unavailable to plants. The greatest areas of boron deficiency are in humid regions where the soils are acid. Every state east of the Mississippi, and generally in the first two rows of states west of the Mississippi, as well as the Pacific states have boron-deficient soils. There are more than 15 functions of boron in plants, most of which are concerned with growth of cells, metabolism, and water relations. The amount of boron needed by any plant is small, but all plants need to absorb boron during their life cycle. Boron is associated with other micronutrients such as Cu, K, and others. The boron for fertilizers comes from Death Valley, the Mojave Desert, and Searl's Lake in California (Russel, 1957).

Molybdenum. Molybdenum is a necessary micronutrient but only a small amount is required, an ounce per acre being more than adequate. Soils low in phosphorus have always been low in Mo. Molybdenum is vital in the life processes of microorganisms, higher plants, and animals. Sandy calcareous soil could not support nitrogen-fixing *Azotobacter* without Mo, and Mo is always essential to the growth of higher plants; this was demonstrated at the University of California in 1940. C. S. Piper at the Waite Agricultural Research Institute near Adelaide, South Australia, demonstrated this need for oats in the same year. At the same institution J. A. Anderson showed by field experiments that the soils there had a low Mo status. The main effect of Mo comes from supplying the nitrogen-fixing bacteria associated with symbiosis in clover with a necessary nutrient. It was also shown that by adding large amounts of lime to the soil Mo could be released in the same way as supplying Mo at the rate of $\frac{1}{16}$ oz (1.8 g) per acre. However, Mo is a primary need of plants as well as a necessity for microorganisms. Some plants are very susceptible to low concentrations of Mo in the soil. The family to which cabbage and cauliflower belong is one such; if a deficiency exists, these plants develop a disease known as whiptail. On the other hand an excess of Mo in soil causes molybdenosis in cattle (teart disease in Somerset, England). Mo deficiencies in soils are caused by weathering and acid fixing. Weathering of soil results in removal of Mo to very low levels, as has been reported for a serpentine soil in California. This soil remains nearly neutral even under heavy rainfall. The Mo released during weathering is leached out and not adsorbed by the clay minerals. An acid may be well supplied with Mo, but the Mo is fixed and not available to plants or microorganisms. Soils in need of Mo are quite limited in the United States; they occur along the Atlantic coast, the eastern part of Washington, and in Hawaii. In Australia and New Zealand Mo deficiency is much more likely to occur.

Sulfur. Plants get sulfur from soil, rain, and irrigation water. Sulfur is an essential micronutrient, and plants absorb about as much sulfur as they do phosphorus. Soil supplies of sulfur are very meager and in some regions deficiencies occur. Industrial areas release sulfur to the atmosphere at the rate of an average of 40 lb/acre/year, and this sulfur is brought to the soil in rainwater (see analyses of rainwater, Table 26, Section 8–2). In nonindustrial areas the rate of sulfur received is only 4 lb/acre/year. One area in southeastern Australia, reported by Hutton (1958), receives practically no sulfur from rain and is consequently sulfur deficient. Data on the sulfate in rainwater from a number of localities in the United States are given by Carroll (1962). Some plants have a particularly high sulfur requirement; for example, the cruciferous (cabbage, etc.), liliaceous, and many flowering plants. Sulfur is well distributed throughout the plant; it is usually absorbed from the soil as sulfate, which is mobile in the sap in the plant. Sulfur may also enter the leaves as SO_2 from the atmosphere. Sulfur is a constituent of all plant proteins, some plant hormones, and amino acids, but the sulfur content of these proteins never exceeds 2%. Sulfur causes protein synthesis, favors the development of root nodules (on legumes), and, consequently, nitrogen fixation. Sulfur is associated with the formation of chlorophyll. Too much sulfur is toxic to plants, an example being the killing of vegetation near smelters. Most of the sulfur in soils of humid regions is in the organic fraction, and it is therefore high in soils that accumulate organic matter. Sulfur is more abundant in the surface soil than in the subsoil, and the concentration of sulfur in some of the great soil groups has been found to be

$$\text{chernozem} > \text{prairie} > \text{red-yellow podzols}$$

Sulfur transformation to organic forms is a microbial process. If the soil is aerated the organic sulfur is oxidized to sulfates which the plants can use, but if the soil is waterlogged the organic sulfur may be reduced to sulfur, sulfureted hydrogen, or other sulfides. One result is the formation of marcasite in bog or marsh soils, and the plants are killed.

Chlorine. Chlorine was proved to be a plant nutrient by T. C. Broyer in 1954 at the University of California. Plants need more chlorine than any of the other micronutrients except iron, and the weight of chlorine required is several thousand times greater than that of molybdenum. Chlorine is widely distributed over the earth's crust by rain. If 10 lb of chlorine were deposited from rainwater per acre, the United States would receive 5 million tons. However, chlorine is not evenly distributed in rainwater because chlorine comes from the sea; rains near the coast contain more chlorine than those inland. Without an oceanic source of chlorine, that in the weathered rocks would soon be depleted. However, because of geographic position some areas receive too much chlorine and its associated sodium, and cyclic salts are a problem in soils. Data on the amounts of chlorine in rainwater are given in Carroll (1962b) and Eriksson (1958, 1959). Because of the wide distribution of chlorine the amount available in natural conditions is probably never limited. The necessity for chlorine was proved in culture experiments, and would be extremely difficult, if not impossible, to prove in field experiments with crops.

Sodium. Sodium is not an essential micronutrient for plants. As it is present in all rocks, and as it is combined as NaCl in rainwater, its occurrence is ubiquitous. It may substitute, to a certain extent, for potassium if this is not readily available. Sodium is merely absorbed along with other chemical elements from the soil water by plants.

Animal Nutrition

Vegetation absorbs the chemical elements that are available in the soils in which it grows; some elements are necessary for life processes, but some are not. The nutrition of animals is founded on the chemical composition of the vegetation in two ways: herbivorous animals utilize the vegetation directly to provide the necessary nutrients except those that are associated with the atmosphere; carnivorous animals utilize the metabolic syntheses of chemical elements that have been made by herbivorous animals. The vegetation is fixed by climate and soil conditions and it concentrates the chemical elements that are available to it. Under natural conditions herbivorous animals have grouped themselves according to climatic and chemical needs; buffalo in the American prairies are an example. Carnivorous animals, although having climatic requirements, are free to move from one area to another and are dependent on the presence of herbivorous animals and other carnivores for their food supply. The result is a more or less balanced plant–animal association in each climatic region.

Most of our knowledge of animal micronutrient requirements has come from studies stemming from the husbandry of herbivorous animals that are required as sources of food for human beings. Much impetus to the study of animal nutrition has come from the change in the land's natural conditions to agricultural conditions in Australia, New Zealand, the United States, and other countries. The endemic animals of these countries had adjusted to the geological, chemical, and pedological conditions but introduced herds and flocks had not, and many micronutritional deficiencies were found. In addition, many farm animals are kept under restricted and unnatural conditions, such as dairy cattle and poultry.

The chemical elements essential for animal growth that have been and are now being studied in animal nutrition are Fe, Cu, Mo, Co, Zn, Se, I, and F. These are essential micronutrients. Other elements that have been found in animals are Al, Si, As, Br, Cd, Cr, Ni, Rb, Sr, Ti, and V. Apart from Sr and V these elements appear to have no useful function, and are ingested because they are present in plants. There is an essential role played by the required micronutrients in animal nutrition, and the following relationships have been established.

Iron. An essential part of the hemoglobin molecule is Fe^{+2}, of which it contains 0.34%. The iron functions as an oxygen carrier by means of oxidation–reduction conditions. Lack of iron produces anemia, a fact that has been known since the time of the Greeks, and probably earlier.

Molybdenum. As noted above Mo is absorbed by plants, and in areas where there is an appreciable amount in the soil cattle develop a molybdenosis which is known in Somerset as "teart." Mo has an inhibiting effect on the ingestion of copper and on its metabolism in animals; this depends on the inorganic-sulfate content of the soil also. Mo is an essential constituent of molybdenoflavoprotein and is necessary for xanthine oxidase in digestion. Calcareous soils with excess Mo cause molybdenosis.

Cobalt. The research that lead to discovering the necessity of cobalt in animal nutrition is one of the highlights of animal research. Its necessity was discovered in 1934 (Underwood, 1962). Cobalt was found as an impurity in limonite that was used to control "coast disease" in New Zealand by Underwood and Filmer (1935). In 1928 it was discovered that the antipernicious anemia factor of liver was a cobalt compound containing 4% Co, and later the factor was found to be vitamin B_{12}. This vitamin is the functional form of cobalt in ruminant metabolism. Research workers at Cornell University in 1951 found that injections of vitamin B_{12} removed all symptoms of Co deficiency in lambs, and it has been shown recently that Co probably plays an essential role in symbiotic nitrogen fixation in legumes; increased growth of clover has resulted from use of Co as a fertilizer. As Co deficiency is actually a deficiency of vitamin B_{12} the real need of Co is in microorganism metabolism in the rumen of cattle and sheep. The microorganisms there synthesize vitamin B_{12}. Excess Co in ruminants causes toxicity. Nonruminant animals have a low requirement of cobalt.

Zinc. In 1869 it was found that zinc is indispensible for the growth of the mold *Aspergillus niger*, but it was not known until 1926 that zinc is also necessary for the growth of higher plants. Zinc is present in all plants and animals and occurs in concentrations similar to iron, and usually in concentrations much greater than those of copper and manganese. Zinc is an essential part of several metalloenzymes. Deficiency of zinc causes malformation of bones and carbohydrate metabolism may be impaired. The amount of zinc in pasture plants growing on normal soils ranges from 30–100 ppm which is sufficient for animal requirements. Zinc deficient soils are unusual, but some such soils have been reported in British Guiana.

Copper. Copper is a necessary dietary component, as copper, in addition to iron, is necessary for the formation of hemoglobin. Copper deficiency manifests itself in many ways in grazing animals: bone formation, pigmentation (a black sheep will produce white wool if Cu deficient), keratinization of wool, and reproduction. There is a close relationship between copper, molybdenum, and inorganic sulfur in metabolism and an antagonism exists between copper and molybdenum. The main storage organ in animals is the liver; most animals contain 10–55 ppm Cu on the dry basis, but sheep and cows contain 100–140 ppm. Copper deficiency is shown by anemia, poor growth, bone disorders, depigmentation of hair or wool, etc. Species of animals differ in their requirements of copper, but its presence is always necessary as it helps in the utilization of iron. However, it was not until 1952, when radioactive iron was used, that it was found that copper increased iron absorption; the necessary amount of copper supplied in soils varies with the geochemical conditions. With high ingestion of calcium carbonate, as in animals grazing on limestone soils, 5 mg/day may be required. In some Western Australian pastures it is considered that 2–4.5 ppm Cu is subnormal, 4.6–7.5 ppm is low normal, 7.5–15 ppm is high normal, and over 15 ppm is Cu-rich. There is a connection between the amount of ferromagnesian minerals in the soil and the copper status (Carroll, 1944). Excess copper is toxic to animals; it may be ingested after application of a copper fertilizer, but would not occur in pastures otherwise.

Selenium. Toxicity due to ingestion of excess selenium that is accumulated by some plants has been known for many years in connection with the soil formed from the Pierre Shale (Cretaceous) in Montana, South Dakota, and Wyoming. This shale averages 18 ppm if the highly selenious Sharon Springs member is included, but the average for the remainder to the shale is only 6 ppm. The Pierre Shale has an outcrop of about 90,000 square miles (Tourtelot, 1962). The selenium comes from the volcanic material which is incorporated in the shale. Murata (1966) estimated that the steam from the pumice cone of Kilauea Iki in Hawaii during the 1959–60 eruption contained 12 ppm Se. Selenium occurs in the soils of Hawaii as it does in Puerto Rico and Iceland. Under high rainfall selenium is leached from the soil and only a nontoxic amount remains. The Pierre Shale, however, is exposed in arid conditions and the selenium is not leached out. It is only very recently that selenium has been found to be an essential micronutrient functionally related to vitamin E. It is an element of considerable importance in the nutrition of many animals, particularly ruminants. It was found by experiment with unthrifty animals that selenium in very minute quantities is a necessary micronutrient. Underwood (1962; p. 292) writes "It is doubtful if there is a comparable example in the history of nutrition of a basic laboratory discovery finding such rapid practical application with results of such wide and immediate significance." At toxic or near-toxic levels selenium is rapidly and efficiently absorbed from naturally seleniferous diets and from soluble salts of selenium. Selenium poisoning of livestocks was reported in the mid-1800's, and it was erroneously named "alkali disease," but later "blind staggers" from Wyoming was found to be similar. Seleniferous soils and vegetation were found to be present in parts of the Great Plains extending up into Canada. Smaller seleniferous areas were found also in Ireland, Israel, northern Australia, and the Soviet Union. Selenium-toxic areas may have such a high content of Se that all the herbage contains selenium, or selenium-accumulator plants may grow in such areas, thereby increasing the selenium intake of animals grazing on these plants. The Se content of plants depends on the species, on the type of soil, and on the climate. Soils containing more than 0.5 ppm Se are regarded as potentially dangerous. Selenium occurs in plants in an inorganic form as selenite or selenate, and in an organic form, replacing the sulfur to a proportion of the normal *S*-containing amino acids. The toxicity of selenium to animals varies with the animal species, the chemical form of selenium, the duration of the intake, and the nature of the diet, especially its protein content. Elemental selenium is relatively nontoxic because of its insolubility, whereas the readily soluble selenates and selenites are quite toxic. Edible herbage on seleniferous areas may contain from 5–20 ppm Se for most plants, but 4000–5000 ppm Se in certain accumulator plants. After numerous experiments selenium therapy was found to be successful in the treatment of a number of animal diseases. In pumice soils in New Zealand selenium is liable to be deficient. Most of the Se-deficient areas are on South Island, but there are also small areas on North Island. The selenium requirements of animals cannot be stated definitely because of a number of variable factors, including the difficulty in determining very small amounts of selenium accurately. On North Island in New Zealand, selenium is required when pastures contain less than 0.03 ppm Se; 0.07–0.14 ppm Se appear to give normal growth. On

South Island, White Muscle Disease occurs with 0.02–0.05 ppm Se. There is apparently a sulfate sulfur–selenium antagonism which complicates natural and experimental assessment of the amount of selenium required.

In dry regions selenium is concentrated by accumulator plants, but in high rainfall regions selenium is leached out of the soil. The amounts of selenium in top soils, largely derived from Cretaceous shales, of dry areas average as follows: the Saskatchewan, 3 ppm; Mexico, 0.275 ppm; Nebraska, 1.76 ppm; South Dakota, 2.34 ppm; Wyoming, 2.05 ppm; Montana, 3.5 ppm; Colorado, 1.37 ppm; Kansas, 2.5 ppm; New Mexico, 2.83 ppm. The soils on Hawaii contain 3.35 ppm Se; Maui, 6.4 ppm; Kauai, 12.75 ppm; all these soils formed on basalt. The soils on Cretaceous shales in Puerto Rico contain 2.23 ppm Se. The latter two groups of soils are in humid climates.

Iodine. Endemic goiter in animals is associated with a deficiency of iodine in food and water. The biochemical process in the utilization of iodine remains obscure; iodine is found in all animal tissues but 70–80% of the iodine is concentrated in the thyroid gland, and all animals require iodine in amounts as follows: poultry, 5–9 mg/day; sheep, 50–100 mg/day; pigs, 80–160 mg/day; cows, 400–800 mg/day. High calcium, as in limestone, and possibly high fluorine, inhibit the absorption of iodine. The main source of iodine is in food, and not in drinking water. Animals fed on pasture grown in an area low in iodine will not receive sufficient iodine in their diet and will be iodine deficient.

Fluorine. Fluorine is an essential micronutrient, and is constantly present in bones and teeth. Although a dietary necessity, excess fluorine is toxic, and causes fluorosis. Fluorine comes from drinking water; plants have only a limited capacity to absorb fluorine. Some rock phosphate used as a fertilizer contains 3–4% fluorine. Both North African and North American rock phosphates are high in fluorine, but not those from the Indian and Pacific Oceans. Chronic fluorosis of sheep and cattle occurs over extensive areas of the United States, Australia, and India, and is primarily a consequence of the consumption of naturally highly fluoridated waters. The highest fluorine is found in waters of deepseated origin; for example, artesian water in Queensland, Australia, commonly contains 2–5 ppm F, and sometimes the it is much higher. Fluorine gets into the bones of animals, which then become anemic and die if not removed to another area where the fluorine content of the water is low. Intermittant dosages of fluorine are less harmful than continued ingestion. Calcium and aluminum salts reduce the toxicity of fluorine.

12–8. Relationship of Trace Elements to Human Nutrition

Trace elements are just as important in human nutrition as they are in animal nutrition, but as the food used is of vegetable and animal origin, and often comes from many different sources, an adequate supply of micronutrients is generally obtained by healthy human beings from normal, well-balanced diets. Humans require the same micronutrients as animals in their diets to maintain health. In many regions of the world, however, the population is dependent on the micronutrients that the local soil, water, and air can supply; in other regions (highly developed countries), food is brought in from many different sources, that is, agricultural areas, and is not dependent on local geochemical and climatic conditions. Furthermore, in these areas livestock may be fed artificially, and so receive balanced diets for productive purposes.

The need for micronutrients became apparent when populations showed symptoms of malnutrition that were traced to a lack of one or more critical elements. Some dietary deficiencies are caused by macronutrients, but a discussion of these is beyond the scope of this book. Trace elements, are necessary for a number of biochemical functions in the body; details of these requirements were discussed for animal nutrition (Section 12–7), which is similar in many respects, although the required amount may not be similar. For example, iron and copper are necessary to the formation and functioning of blood, cobalt is required for vitamin B_{12}, and selenium for vitamin E.

Deficiencies of micronutrients in human nutrition have not been recorded for Co, Cu, Mn, Mo, Se, and Zn. The intake of excessive amounts of many micronutrients results in toxicity that is shown by impaired growth, anemia, and susceptibility to disease; however, most toxicity in human beings has resulted from industrial hazards, and not from micronutrients supplied by diets. But it is becoming increasingly evident that poor public health in certain regions may be due to the biogeochemistry of the region (Cannon and Davidson, 1967).

It has been known for a very long time that iron is necessary in human diets, and that anemia results from lack of iron, which, however, is present in appreciable amounts in most rocks (5.6×10^4 ppm in the earth's crust) and is determined in routine chemical analyses of rocks and minerals. Two elements that are critical in human metabolism are *iodine*, which is necessary for the correct functioning of the thyroid gland, and *fluorine*, which occurs in teeth and bones. The amount of fluorine in the human and animal diet ranges from essential to toxic within rather narrow limits. The source of iodine is the world ocean, and iodine is carried to the land surfaces by rain. Therefore the distribution of iodine is largely controlled by the access of rain to the land, and by the rainfall pattern. After precipitation by rain, iodine is adsorbed on the clay and organic matter in soils. The amount of iodine in sea water is 0.05 mg/liter (Goldberg, 1957). There are very few figures for the amount of iodine in soils, and iodine is generally not determined in the chemical analysis of rainwater. Vinogradov (1959) gave figures for the amount of iodine in the soils of the Russian Platform ranging from 1–12 ppm for the surface soils and less for the subsoils. The amount of iodine in these soils shows a slight trend in the direction of podzolic soils containing less than chernozems, both of which contain less than desert soils; but the number of iodine determinations is small, and other factors, such as rainfall, may control the distribution of iodine. Swaine (1955) gives no figures for iodine. Mitchell (1964) states that although the average iodine content of rocks (more in granitic than in basaltic rocks) is 0.3 ppm, the normal range for soils is 0.6–8.0 ppm. Shacklette and Cuthbert (1967) point out that iodine is derived principally from the oceans, and the content in soils is not dependent on the iodine content of the rocks from which the soils were derived. The main source of iodine for both humans and animals is in the food consumed and is partly controlled by the water utilized by plants for growth. Lack of iodine has been known since the time of the Greek culture to cause goiter; endemic goiter is associated primarily with a deficiency of iodine. In areas where iodine is likely to be deficient, iodine is added to the salt used domestically. The amount of iodine required in humans is 100–200 μg daily (Underwood, 1962). Public health authorities have recorded numerous areas in the world that are iodine deficient in which endemic goiter may occur (Underwood, 1962; p. 221). All land plants contain iodine but plant species vary in their ability to take in iodine. Marine plants contain much more iodine than land plants, and their use (together with marine animal products) has been an important influence in controlling goiter in countries which would otherwise show a high incidence of goiter. Calcareous rocks, such as limestone, inhibit the intake of iodine, and it is possible that

TABLE 37

Rock	F range, ppm	F average, ppm
Granite	0–3.4	0.9
Basalt	0–0.4	0.2
Andesite	0–0.1	0.1
Limestone	0–0.9	0.3
Dolomite	0–1.7	0.5
Sandstone and quartzite	0–1.9	0.4
Clay	0–2.8	0.6

calcareous soils have the same effect. Calcareous soils in dry areas may be suspected as inhibiting the absorption of iodine by plants, but irrigation may ameliorate this condition. On the other hand, podzolic soils are leached rather rapidly and lose any iodine that is received in rain, as iodine is quite soluble and easily removed. The principal commercial sources of iodine are Chilean saltpeter and the ashes of seaweeds. Prior to the use and processing of Chilean nitrate deposits iodine was recovered entirely from the ashes of seaweeds.

Excess fluorine in the groundwater causes fluorosis in animals (anaemia, leisons in bones and teeth, and eventual death) and mottling of tooth enamel in humans. Fleischer (1967) calculated the range of fluorine in waters associated with various kinds of rocks in the United States (Table 37).

Analyses of groundwater in the coterminous United States were plotted on a map to show the range of fluorine content (Fleischer, 1967). The four ranges used in this compilation were 1.5 ppm F or higher, 1.0–1.4 ppm F, 0.5–0.9 ppm F, and 0–0.4 ppm F. Extensive areas in the southwestern and western United States have the highest fluorine contents, and areas from the eastern Great Plains to the Atlantic Coast have the lowest fluorine contents; however, the distribution is not uniform and is a consequence of the rock formation and rainfall. North American diets contain 0.3–0.5 mg daily; most foods contain less than 0.5 ppm F. Dental fluorosis (mottled teeth) is caused by continuous intake of near-toxic or toxic amounts of fluorine daily. Water containing less than 1 ppm F is not toxic. Studies of the incidence of dental caries (Underwood, 1962) show that 1 ppm F in water gives maximum health with maximum safety. Fluoridation of water to maintain 1.0–1.2 ppm F in the United States is beneficial to tooth development and reduces dental caries. Table 38 lists the effects on tooth formation caused by differents of fluorine in drinking water.

TABLE 38

F, ppm	Effect
1–2	Mild mottling of enamel
2–4	Mottling, staining, and erosion
4–6	Deep staining, mottling, and pitting
6	Teeth badly affected
8	Skeletal fluorosis occurs

Appendix 1

MINERAL TRANSFORMATIONS IN WEATHERING

Fresh rock adjusting to equilibrium conditions on the earth's surface comes into contact with air and with water of a certain composition. When both air and water can penetrate the rock the following changes take place.

The ferrous iron in minerals oxidizes to ferric iron. In place of the Fe^{+2} ion in a structural relationship to other ions in the mineral, ferric iron is concentrated in cracks in and between mineral grains. This ferric iron is generally termed "limonite." It may be amorphous ferric oxide or it may be goethite that has crystallized from the amorphous iron oxide. This iron oxide may adhere to the outside of other minerals imparting a brownish color to them. Plate 1 shows the coloration produced in saprolites by iron.

Aluminum in feldspar is attacked by water that has a pH either below pH 4 or above pH 9. The attack is not direct. Nash and Marshall (1956) have shown that the original alteration of feldspar is a cation-exchange reaction, whereby the alkali and alkaline-earth ions are removed (K, Na, Ca). The exchange is a surface reaction. The crystalline structure of the feldspar grain is destroyed, and a thin zone of altered material consisting of alumina and silica results.

The grain becomes carious from the outside, and it is in the carious zone that reaction with water occurs. Goldich's (1938) reaction series of mineral stability is based on the observed sequence of alteration of the common rock-forming minerals. The alterations that have been observed (Garrels, 1957) are

K-feldspar → K-mica → Kaolinite → Gibbsite

$$\left.\begin{matrix}\text{Na}\\ \\ \text{Ca}\end{matrix}\right\}\text{-feldspar} \rightarrow \text{Montmorillonite} \rightarrow \text{Kaolinite} \rightarrow \text{Gibbsite}$$

In a system open to rainwater (a steady-state chemical environment) the rock-forming silicates tend to alter to last-formed aluminum hydrates (and ferric oxide where ferric minerals are originally present). The water at the stage of aluminum removal is alkaline as a result of hydrolysis, which can be thought of as the abrasion pH (Chapter 8). All of the rock-forming minerals except quartz have alkaline abrasion-pH values. When the interstititial water of a rock reaches pH 8–9 (Fig. 23, Section 8–3) aluminum goes into solution, and first boehmite ($Al_2O_3 \cdot 2H_2O$) and then gibbsite ($Al_2O_3 \cdot 3H_2O$) is formed. Harrison (1934) described the early formation of gibbsite in the weathering of granite in British Guiana.

In the crystal structure changes in feldspar alteration, feldspar with frameworks of linked SiO_2 tetrahedra and Al_2O_3 octahedra having the cations K, Na, Ca, and Ba situated in the interstices of the negatively charged tetrahedra reverts to simple silica tetrahedra and alumina octahedra when these cations are removed (in the order Ca $>$ Na $>$ K). In micas (fine white mica or sericite is commonly recognized as an alteration of feldspar) the K^+ ion has been removed from its interstitial position in a silica tetrahedron and is now between the composite layers of silica tetrahedra and alumina octahedra, where it binds two layers together. This is the common structure of all micas. When the K^+ ion is removed from its central binding position the two composite layers become independent and kaolinite results. As silica is removed during alteration, the early-formed aluminum hydrates are silicated to form additional kaolinite.

In a rock that contains nepheline ($Na_6K_2Al_8Si_9O_{34}$) instead of feldspar, alteration is more rapid. Nepheline gelatinizes with acid solutions (Murata, 1943), whereas there is a surface exchange reaction with feldspars. The abrasion pH of feldspar is 9–10, but that of nepheline is 11. Because the nepheline structure breaks down completely with acid solutions the Al^{+3} ion is readily mobilized and at pH 11 is in solution. With the release of alumina, silica can become mobilized and removed from its original position in the nepheline crystal structure. Gibbsite forms readily in weathering nepheline syenite, and bauxite results as in the Arkansas bauxite region (Gordon et al, 1958). Other examples are the gibbsite formation in the weathering of the nepheline syenite of the Khibny Tundra in Russia (Dorfman, 1958), and nepheline syenite of the Iles de Los, western Africa (Millot and Bonifas, 1955).

Although gibbsite forms on the weathering of both feldspar- and nepheline-bearing rocks, the process differs, and the weathering of nepheline-bearing rocks is more rapid than that of feldspar-bearing rocks.

Si^{+4} in minerals other than quartz is slightly soluble (about 140 ppm) at all pH values (Fig. 23, Chapter 8); it is slightly more soluble at higher pH values. It is the soluble SiO_2 that combines with aluminum hydroxide to form kaolinite. As a result of these changes the rock has become a saprolite which consists of:

Quartz: if present, remains largely unaltered. The grain size may be the same as in the unaltered rock, but probably the size has been reduced by splitting along cracks.

Feldspar: changed to mica and/or kaolinite.

Ferromagnesian minerals: alteration starts by the oxidation of Fe^{+2} to Fe^{+3}, where exposed at the edges of crystals. A second alteration results in the change of optical properties caused by the removal of cations and consequent change in structure. Pyroxene has a less stable structure than amphibole. The metamorphic types of amphibole are very stable minerals. Micas lose the K^+ ion binding the layers together, and the muscovite micas adsorb H^+ and other ions thereby reverting to degraded forms and illite-like minerals. Biotite micas are much less stable and tend to disappear as discrete grains, although they form part of the chlorite-type minerals commonly present in weathered products.

The alterations tend to follow the stability series as given by Goldich (1938). This saprolitic material, altered to a greater or lesser extent, forms the C horizon for the continuation of the weathering process that results in soils. Garrels (1957) has described the alterations in weathering thermodynamically. He visualizes a simplified, idealized system as "a vertical soil profile under conditions of high rainfall and continuous downward drainage as a steady-state condition with zones of stability of the minerals from unaltered silicates at the bottom to aluminous residue near the top." Figure 15 (Chapter 6) shows this to be true for the mineralogy of the soils of Guam. However, it is preferable first to gain some knowledge of the weathering process in saprolite formation. Soil development has a great many variables that contribute to its nature. An example of diversification of soils of one type of parent rock is given in Plate 3.

Minerals exist as electrically neutral entities, but they have inherent energies that are known as free energies; this is the maximum available work that can be obtained in going from the initial to the final state. Garrels (1957; p. 788) has calculated a number of standard free energies of reactions and of formation of compounds (Table 39).

Mineral transformations that take place in soil formation are imposed on those caused by geochemical weathering. The minerals that were derived by weathering of rock minerals are further modified, depending on their stability, in pedochemical weathering. The distinguishing features of soil profiles are due to chemical and mineralogical changes (Chapters 3 and 4). Chemical weathering of minerals in soils has been described by Jackson

TABLE 39

Standard Free Energies (after Garrels, 1957)

Reaction	Value, kcal at 25°C
$Al_2O_3 \cdot 3H_2O_c + 2SiO_{2\,aq} = H_4Al_2Si_2O_9 + H_2O_{liq}$	−10.3
$Al_2O_3 \cdot 3H_2O_c + 2SiO_{2\,glass} = H_4Al_2Si_2O_{9c} + H_2O_{liq}$	−17.3
$Al_2O_3 \cdot 3H_2O_c + 2SiO_{2\,quartz} = H_4Al_2Si_2O_{9c} + H_2O_{liq}$	−20.5
$3H_4Al_2Si_2O_{9c} + 2K^+_{aq} = 2KAl_3Si_3O_{10}(OH)_{2c} + 3H_2O_{liq} + 2H^+_{aq}$	+18
$KAl_3Si_3O_{10}(OH)_{2c} + 2K^+_{aq} + 6SiO_{2\,quartz} = 3KAlSi_3O_{8c} + 2H^+_{aq}$	+20
Mineral	
Kaolinite	−883
K-mica	−1298
K-feldspar	−856

All reaction values estimated ±2 kcal, with strong possibility of larger errors in values for minerals.

and Sherman (1953). A sequence of thirteen stages in weathering has been recognized in soils from various areas. Local as well as regional chemical environments are responsible for this mineralogical sequence that applied particularly to clay-sized particles. The sequence is as follows:

1. Gypsum (halite, $NaNO_3$, NH_4Cl, other salts)
2. Calcite (dolomite, aragonite, apatite)
3. Olivine (pyroxenes, amphiboles)
4. Biotite (glauconite, chlorites, nontronite)
5. Albite (anorthoclase, stilbite, orthoclase, microcline*)
6. Quartz (cristobalite, opal)
7. Muscovite (sericite, illite)
8. Interstratified 2:1 layer silicates and vermiculite (including partially expanded hydrous micas, randomly interstratified 2:1 layer silicates, and regularly interstratified 2:1 layer silicates)
9. Montmorillonite (and the members of the montmorillonite group)
10. Kaolinite (halloysite, disordered kaolinite)
11. Gibbsite (boehmite, allophane, etc.)
12. Hematite (goethite and other iron oxides)
13. Zircon (ilmenite, rutile, anatase, leucoxene, tourmaline, metamorphic aluminosilicates)

These stages are similar to those described for sand-sized grains in

* Microcline is much more resistant to weathering than orthoclase or any of the other feldspars.

Chapter 4. The weathering sequence of vitreous volcanic ash reported from many areas that have been subjected to sufficient rainfall for a considerable length of time is

$$\text{volcanic glass} \rightarrow \text{allophane} \rightarrow \text{(halloysite) kaolinite}$$

In general, the sequence of mineralogical changes follows that given in Fig. 9 (Section 4–4).

Appendix 2

DATA OF ROCK WEATHERING*

Geologists and soil scientists have made some 350–360 detailed studies of the alteration of individual rocks into soils by weathering *in situ* (Table 40). Not all these investigations give as many mineralogical, chemical, and physical details about the weathering products as would be desirable, but each description gives a great deal of information about the rock from a particular locality. The weathering localities have been classified according to Köppen's (1931) climatic zones. The intensity of the weathering in these zones is:

$$A >> B > C > D >>> E$$

* Publication authorized by the Director, U.S. Geological Survey.

TABLE 40
Classification of Data of Weathering Rocks According to Rock Type, Climate, and Country (Locality)

Rock type	Climates (Köppen classification)				
	A Tropical (Aw''a, Aw, Awa, ama, aw''h, afh, am, afa)	B Dry (Bwk, Bwh, Bsk, Bsh, Bfh)	C Humid, mesothermal (Csa, Csb, Cfe, Cfh, Cwa)	D Humid, microthermal (Df, Dfa, Dfb, Dfc, Dwe)	E Polar (E, ESc)
Granite	*British Guiana* (Harrison, 1934) *India* (Fox, 1936) *Malaya* (Blanck, 1949; Blanck, Credner, and von Oldershausen, 1935)	*Egypt* (Blanck, 1949)	*Australia* (Carroll and Jones, 1947) *Chile* (Blanck, 1949; Blanck, Reiser, and Oldershausen, 1933) *England* (Butler, 1953) *Germany* (Preusse, 1957) *United States* Southern U.S. (Harriss and Adams, 1966) Georgia (Merrill, 1913) Missouri (Humbert and Marshall, 1943) Wyoming (Short, 1961)	*USSR* (Glinka, 1935)	*Spitzbergen* ([illegible] 1949)

Granodiorite			*Australia* (Brewer, 1955) *Japan* (Yamasaki, Iida, and Yokoi, 1955)	
Porphyry				*Japan* (Suwa and others, 1958)
Pegmatite	*British Guiana* (Harrison, 1934)	*Egypt* (Blanck, 1949)		
Gneiss (all types)	*Ceylon* (Pannebokke, 1959) *Ghana* (Stephen, 1953) *India* (Fox, 1936; Harrassowitz, 1926; Mohr and Van Baren, 1954) *Tanganyika* (Muir, Anderson, and Stephen, 1957)	*India* (Tamhane and Namjoshi, 1959)	*Australia* (Carroll and Jones, 1947; Costin, Hallsworth, and Woof, 1952) *England* (Butler, 1953) *Norway* (Blanck, 1949) *United States* District of Columbia (Merrill, 1913) Georgia (England and Perkins, 1959) Virginia (Merrill, 1913)	*Germany* (Blanck and Melville, 1940) *Switzerland* (Blanck, 1949) *United States* Minnesota (Goldich, 1938) New Jersey (Connor, Shimp, and Tedrow, 1957)

TABLE 40 (continued)

Rock type	Climates (Köppen classification)				
	A Tropical	B Dry	C Humid, mesothermal	D Humid, microthermal	E Polar
Syenite and monzonite	*French West Africa* (Millot and Bonifas, 1955) *Thailand* (Blanck, Credner, and Oldershausen, 1935)		*United States* Arkansas (Merrill, 1913; Gordon, Tracey and Ellis, 1958)	*Norway* (Butler, 1954)	*USSR* (Vinogradov, 1959)
Gabbro and metagabbro			*England* (Butler, 1953) *Scotland* (Glentworth, 1944) *Syria* (Reifenberg, 1952) *United States* Maryland (Carroll, 1953) North Carolina (Hardy and Rodrigues, 1939b; Middleton, 1930)		

Norite and gabbro-norite	*Sierra Leone* (Martin and Doyne, 1927)		*Cyprus* (Reifenberg and Ewbank, 1933II) *South Africa* (Lombaard, 1934)	
Andesite, dacite, meta-andesite	*Grenada, BWI* (Hardy and Rodrigues, 1939a) *Indonesia* (Blanck, 1949)		*Australia* (Anderxon, 1941) *Chile* (Blanck, 1949) *United States* North Carolina (Short, 1961)	
Basalt and basaltic lava	*Hawaii* (Abbott, 1958; Cline et al., 1955; Hough and Byers, 1937; Sherman and Uehara, 1956; Hough, Gile, and Foster, 1941; Sherman, 1955) *Samoa* (Seelye, Grange, and Davis, 1938) *Guam* (Carroll and Hathaway, 1963)	*India* (Tamhane and Namjoshi, 1959) *United States* Colorado (Short, 1961) *USSR* (Vinogradov, 1959; Glinka, 1935)	*Australia* (Raggatt, Owen, and Hills, 1945; Tiller, 1959; Hanlon, 1944; Carroll and Woof, 1951) *Cyprus* (Reifenberg and Ewbank, 1933II) *Germany* (Preusse, 1957; Huffmann, 1954; Blanck and Melville, 1940) *Northern Ireland* (Brown, 1954; Eyles, 1952; Charlesworth and others, 1935) *Italy* (Blanck and Melville, 1940) *Lebanon*(Reifen-berg, 1952)	*Iceland* (Valek, 1959) *Norway* (Butler, 1954)

TABLE 40 (continued)

Rock type	Climates (Köppen classification) A Tropical	B Dry	C Humid, mesothermal	D Humid, microthermal	E Polar
Diabase and dolerite	*British Guiana* (Harrison, 1934) *Guiana* (Blanck, 1949) *Natal* (Beater, 1947)	*Rhodesia* (Ellis, 1952)	*Australia* (Tiller, 1957, 1959) *Cyprus* (Reifenberg and Ewbank, 1933II) *United States* Missouri (Humbert and Marshall, 1943) North Carolina (Hardy and Rodrigues, 1939b; Middleton, 1930) Virginia (Merrill, 1913)	*United States* Mass. (Merrill, 1913; Goldich, 1938) Minn. (Goldich, *USSR* (*Makeyev* and Nogina, 1958)	
Diorite and epidiorite	*British Guiana* (Harrison, 1934)		*Chile* (Blanck, 1949) *England* (Stephen, 1952II) *Germany* (Preusse, 1957) *United States* North Carolina (Cady, 1951) Virginia (Merrill, 1913)	*USSR* (Glinka, 1935)	*Antarctica* (Kelly and Zumberge, 1961)

Amphibolite	*Congo* (Waegemans, 1954) *Caroline Islands* (Blanck, 1949)		*United States* S. Dakota (Goldich, 1938)
Dunite	*French West Africa* (Millot and Bonifas, 1955)		
Serpentine	*British Solomon Islands* (Birrell, Seelye, and Grange, 1939) *Cuba* (Bennett and Allison, 1928; Robinson, Edgington, and Byers, 1935) *New Caledonia* (Birrell and Wright, 1945) *Puerto Rico* (Roberts, 1942; Mohr and van Baren, 1954)	*Cyprus* (Reifenberg and Ewbank, 1933I) *England* (Butler, 1953) *Syria* (Reifenberg, 1952) *United States* Maryland (Robinson, Edgington, and Byers, 1935) Virginia (Robinson, Edgington, and Byers, 1935) Oregon (Robinson, Edgington, and Byers, 1935)	

TABLE 40 (continued)

Rock type	Climates (Köppen classification)				
	A Tropical	B Dry	C Humid, mesothermal	D Humid, microthermal	E Polar
Talc			*United States* Virginia (Merrill, 1913; Robinson, Edgington, and Byers, 1935)		
Schists (horn-blende, muscovite, biotite, clay)	*Bolivia* (Blanck, Reiser, and Oldershausen, 1933) *British Guiana* (Harrison, 1934) *Ghana* (Stephen, 1953)		*England* (Stephen, 1952I) *Germany* (Preusse, 1957) *United States* Maryland (Merrill, 1913) Virginia (Rich and Obenshain, 1955)	*Norway* (Butler, 1954; Holtedahl, 1953) *Switzerland* (Blanck, 1949)	*Spitzbergen* (Blanck, 1949; Smith, 1956)
Muscovite	*British Guiana* (Harrison, 1934)				

Tuff and agglomerate	*Guam* (Carroll and Hathaway, 1963) *Norfolk Island* (Stephens and Hutton, 1954; Hutton and Stephens, 1956) *Puerto Rico* (Bonnet, 1939)		*Australia* (Tiller, 1958) *Columbia* (Barshad and Rojas-Cruz, 1950)	*Iceland* (Válek, 1959)	
Volcanic ash	*St. Vincent, BWI* (Hay, 1960)		*Japan* (Kamoshita, 1958)	*United States* California (Dickson and Crocker, 1953)	
Pumice	*Indonesia* (van Baren, 1931)				
Sandstone, quartzite	*Thailand* (Blanck, Credner and Oldershausen, 1935)	*Bolivia* (Blanck, 1949)	*Cyprus* (Reifenberg and Ewbank, 1933I) *Germany* (Blanck and Melville, 1940); Blanck, 1949; Preusse, 1957; Huffmann, 1954) *Lebanon* (Reifenberg, 1952)	*Switzerland* (Blanck, 1949) *United States* Pennsylvania (Connor, Shimp, and Tedrow, 1957) Wisconsin (Wurman, 1960)	*Spitzbergen* (Blanck, 1949)

TABLE 40 (continued)

Rock type	Climates (Köppen classification)				
	A Tropical	B Dry	C Humid, mesothermal	D Humid, microthermal	E Polar
Arkose			*New Zealand* (McLaughlin, 1955)		
Shales, claystone	*India* (Tamhane and Namjoshi, 1959) *Malaya* (Alexander, 1959) *Sumatra* (Blanck, 1949) *Trinidad* (Rodrigues and Hardy, 1947)			*United States* New Jersey (Connor, Shimp, and Tedrow, 1957)	
Graywacke		*Egypt* (Blanck, 1949)	*Germany* (Preusse, 1957)		

Limestone and dolomite	*Guam* (Carroll and Hathaway, 1963) *Indonesia* (van Baren, 1930) *Malaya* (Blanck, Credner, and Oldershausen, 1935) *Marshall Islands* (Fosberg, 1953; Fosberg and Carroll, 1965) *Savage Island* (Birrell, Seelye, and Grange, 1939) *Thailand* (Blanck, Credner, and Oldershausen, 1935)	*India* (Tamhane and Namjoshi, 1959	*Cyprus* (Reifenberg and Ewbank, 1933I) *Germany* (Preusse, 1957; Huffmann, 1954; Blanck, 1949) *Holland* (van Baren, 1930) *United States* Arkansas (Merrill, 1913) Missouri (Brydon and Marshall, 1935) New Jersey (Connor, Shimp, and Tedrow, 1957; Krebs and Tedrow, 1957) Texas (Nelson, Kunze, and Godfrey, 1960) Virginia (Merrill, 1913; Carroll and Hathaway, 1954)	*Switzerland* (Blanck, 1949)

TABLE 40 (continued)

Rock type	Climates (Köppen classification)				
	A Tropical	B Dry	C Humid, mesothermal	D Humid, microthermal	E Polar
Glacial sediments (till, boulder clay, varve clay)			*Scotland* (Glentworth, 1944)	*USSR* (Vinogradov, 1959; Zaboyeva, 1958)	*Antarctica* (Blakemore and Swindale, 1958) *Norway* (Holteda hl, 1953) *United States* Alaska (Drew and Tedrow, 1957)
Loess			*Germany* (Preusse, 1957) *United States* Illinois (Whiteside and Marshall, 1944) Missouri (Whiteside and Marshall, 1944)	*Iceland* (Válek, 1959)	
Phosphate rock			*United States* Florida (Altschuler, Jaffe, and Cuttita, 1956)		

Greensand		*Australia* (Hosking and Greaves, 1936; Cole, 1943)	
Deltaic and other sediments	*USSR* (Vinogradov, 1959)	*United States* Texas (Kunze and Oakes, 1957)	
Alluvium, colluvium, undifferentiated	*USSR* (Vinogradov, 1959; Popov, 1958)		*USSR* (Vinogradov, 1959)
Desert "meal"	*Chile* (Blanck, 1949)		

LITERATURE CITED

Abbott, A. T., 1958, Occurrence of gibbsite on the island of Kauai, Hawaiian Islands, *Econ. Geology*, **53**, 842–855.

Alexander, F. E. S., 1959, Observations on tropical weathering; a study of the movement of iron, aluminium, and silicon in weathering rocks at Singapore, *Geol. Soc. London Quart. Jour.*, **115**, 123–144.

Altschuler, Z. S., Jaffe, E. B., and Cuttita, F., 1956, The aluminum phosphate zone of the Bone Valley formation, Florida, and its uranium deposits, *in:* Contributions to the geology of uranium and thorium by the United States Geological Survey and Atomic Energy Commission for the United Nations international conference on peaceful uses of atomic energy, Geneva, 1955, L. R. Page *et al.*, *U.S. Geol. Survey Prof. Paper*, No. 300, pp. 495–504.

Anderson, V. G., 1941, The origin of the dissolved inorganic solids in natural waters with special reference to the O'Shannassy River catchment, Victoria, *Australian Chem. Institute Jour. & Proc.*, **8**, 130–150.

Anderson, D. H., and Hawkes, H. E., 1958, Relative solubility of the common elements in weathering of some schist and granite areas, *Geochim. Cosmochim. Acta*, **14**, 204–211.

Atkinson, H. J., and Wright, J. R., 1957, Chelation and the vertical movement of soil constituents, *Soil Science*, **84**, 1–11.

Baas-Becking, L. G. M., Kaplan, I. R., and Moore, D., 1960, Limits of the natural environment in terms of pH and oxidation–reduction potentials, *Jour. Geology*, **68**, 243–284.

Barshad, I., 1964, Chemistry of soil development, *in: Chemistry of the Soil*, 2nd ed., Firmin E. Bear (editor): Reinhold Publishing Corp., New York, pp. 1–70.

Barshad, I., and Rojas-Cruz, L. A., 1950, A pedologic study of a podzol soil profile from the equatorial region of Columbia, South America, *Soil Science*, **70**, 221–236.

Barth, T. W., 1948, Oxygen in rocks: a basis for petrographic calculations, *Jour. Geology*, **56**, 50–61.

Bates, T. F., 1960, Rock weathering and clay formation in Hawaii, *Mineral Ind.*, **20**(8), 4–6.

Baver, L. D., 1956, *Soil Physics*, 3rd ed.: John Wiley & Sons, New York.

Beater, B. E., 1947, Chemical composition of some Natal coastal dolerites and their alteration products, *Soil Science*, **64**, 58–96.

Becker, G. F., 1895, A reconnaissance of the goldfields of the southern Appalachians, *U.S. Geological Survey, 16th Annual Report*, Pt. 3, pp. 251–331.

Benes, N. S., 1960, Soil temperatures at Cape Hallett, Antarctica, 1958, *Monthly Weather Review*, **88**, 223–227.

Bennett, H. H., and Allison, R. V., 1928, *The Soils of Cuba*, Tropical Research Foundation, 410 pp.

Birrell, K. S., and Wright, L. D., 1945, A serpentine soil in New Caledonia, *New Zealand Jour. Sci. Technology*, **27 A,** 72–76.

Birrell, K. S., Seelye, F. T., and Grange, L. I., 1939, Chromium in soils of western Samoa and Niue Island, *New Zealand Jour. Sci. Technology*, **21 A,** 91–95.

Blakemore, L. C., and Swindale, L. D., 1958, Chemistry and clay mineralogy of a soil sample from Antarctica, *Nature*, **182,** 47–48.

Blanck, Edwin, 1949, *Einführung in die genetische Bodenlehre als selbstständige Naturwissenschaft und ihre Grundlagen*, Göttingen, Vandenhoeck und Ruprecht, 420 pp.

Blanck, E., and Melville, R., 1940, Untersuchungen über die rezente und fossile Verwitterung des Gesteins innerhalb Deutschlands, etc., *Chemie der Erde*, **13**(3), 99–201; 235–315; 377–471.

Blanck, E., and Melville, R., 1942, Untersuchungen über die rezente und fossile Verwitterung des Gesteins innerhalb Deutschlands, etc. (continued), *Chemie der Erde*, **14,** 1–106, 253–311, 386–452.

Blanck, E., Credner, W., and von Oldershausen, E., 1935, Beiträge zur chemischen Verwitterung und Bodenbildung in Siam, *Chemie der Erde*, **9,** 422.

Blanck, E., Reiser, A., and von Oldershausen, E., 1933, Beiträge zur chemischer Verwitterung und Bodenbildung Chiles, *Chemie der Erde*, **8,** 339–439.

Bloomfield, C., 1953–54, A study of podzolization, 1–1V: *Jour. Soil Sci.*, **4–5.**

Bonnet, J. A., 1939, The nature of lateritization as revealed by chemical, physical, and mineralogical studies of a lateritic soil from Puerto Rico, *Soil Science*, **48,** 25–40.

Brenmer, J. N., 1951, 1954, A review of recent work on soil organic matter. I, *Jour. Soil Sci.*, **2,** 67–82; II, *ibid.*, **5,** 214–232.

Brewer, Roy, 1955, Mineralogical examination of a yellow podzolic soil on granodiorite, *Commonwealth Sci. Indust. Research Organization (Australia) Soil Publ.*, No. 5, 28 pp.

Brindley, G. W., Bailey, S. W., Faust, G. T., Forman, S. A., and Rich, C. I., 1968, Report of the Nomenclature Committee (1966–67) of the Clay Minerals Society, *Clays and Clay Minerals*, **16**(4), 322–324.

Broadbent, F. E., 1957, Soil organic matter–metal complexes. II. Cation exchange chromatography of copper and calcium complexes, *Soil Sci.*, **84,** 127–131.

Broadbent, F. E., 1962, Biological and chemical aspects of mineralisation, *Internat. Soc. Soil Sci., Trans. Joint Meeting Commissions 4 and 5*, Soil Bureau, Lower Hutt, New Zealand, pp. 220–229.

Brown, G., 1961, *The X-ray Identification and Crystal Structures of Clay Minerals*, The Mineralogical Society, London.

Brown, W. O., 1954, Some soil formations of the basaltic region of northeast Ireland, *Irish Naturalists' Jour.*, **11,** 120–132.

Bryan, W. H., 1939, The red earth residuals and their significance in southeastern Queensland, *Royal Soc. Queensland Proc.*, **50,** 21–32.

Brydon, J. E., and Marshall, C. E., 1935, Mineralogy and chemistry of the Hagerstown soil in Missouri, *Missouri Univ. Agric. Expt. Sta. Research Bull.*, No. 655, 56 pp.

Butler, B. E., 1956, Parna—an aeolian clay, *Australian Jour. Sci.*, **18,** 145–151.

Butler, J. R., 1953, The geochemistry and mineralogy of rock weathering. I. The Lizard area, Cornwall, *Geochim. Cosmochim. Acta*, **4**(4), 157–178; The geochemistry and mineralogy of rock weathering. II. The Nordmarka area, Oslo, *Geochim. Cosmochim. Acta*, **6**(5–6), 268–281.

Buchanan, F., see Fox, C. S.

Cady, J. G., 1951, Rock weathering and soil formation in the North Carolina Piedmont region, *Soil Sci. Soc. America Proc.*, **15,** 337–342.

Campbell, J. M., 1917, Laterite: Its origin, structure, and minerals, *Mining Magazine*, **17,** 17–77, 120–128, 171–178, 220–228.

Cannon, H. L., 1960, Botanical prospecting for ore deposits, *Science*, **132**(3427), 591.

Cannon, H. L., and Davidson, D. F., 1967, Relation of geology and trace elements to nutrition, *Geol. Soc. America Special Paper*, No. 90, 68 pp.

Cannon, H. L., and Kleinhampl, F. J., 1956, Botanical methods of prospecting for uranium, *U.S. Geol. Survey Prof. Paper*, No. 300, pp. 681–686.

Carroll, Dorothy, 1939, Sand-plain soils of the Yilgarn Goldfield, *Geol. Survey Western Australia Bull.*, No. 97, pp. 161–180.

Carroll, Dorothy, 1944, Mineralogical examination of some soils from southwestern Australia, *Jour. Dept. Agric. Western Australia*, **21 (n.s.)**, 82–93.

Carroll, Dorothy, 1945, Mineralogy of some soils from Denmark, Western Australia, *Soil Science*, **60**, 413–426.

Carroll, Dorothy, 1953, Description of a monalto soil in Maryland, *Soil Science*, **76**, 87–102.

Carroll, Dorothy, 1958, Role of clay minerals in the transportation of iron, *Geochim. Cosmochim. Acta.*, **14**, 1–27.

Carroll, Dorothy, 1959a, Sedimentary studies in the Middle River drainage basin of the Shenandoah Valley of Virginia, *U.S. Geol. Survey Prof. Paper*, No. 314–F, 19 pp.

Carroll, Dorothy, 1959b, Ion exchange in clays and other minerals, *Geol. Soc. America Bull.*, **70**, 749–780.

Carroll, Dorothy, 1960, Ilmenite alteration under reducing conditions in unconsolidated sediments, *Econ. Geology*, **55**, 618–619.

Carroll, Dorothy, 1962, Rainwater as a chemical agent of geologic processes, *U.S. Geol. Survey Water-Supply Paper*, No. 1535–G, 18 pp.

Carroll, Dorothy, and Hathaway, J. C., 1954, Clay minerals in a limestone soil profile (Va.), *in: Clay and Clay Minerals*, Swineford and Plummer (eds.), Natl. Res. Council Publ. 327, pp. 171–182.

Carroll, Dorothy, and Hathaway, J. C., 1963, Mineralogy of selected soils from Guam, *U.S. Geol. Survey Prof. Paper*, No. 403–F, 53 pp.

Carroll, Dorothy, and Jones, N. K., 1947, Laterite developed on acid rocks in southwestern Australia, *Soil Science*, **64**, 1–15.

Carroll, Dorothy, and Starkey, H. C., 1959, Leaching of clay minerals in a limestone environment, *Geochim. Cosmochim. Acta*, **16**, 83–87.

Carroll, Dorothy, and Woof, Marion, 1951, Laterite developed on basalt at Inverall, New South Wales, *Soil Science*, **72**, 87–99.

Carter, G. F., and Pendleton, R. L., 1956, The humid soil: Process and time, *Geograph. Review*, **46**, 488–507.

Charlesworth, J. K., *et al.*, 1935, *The Geology of Northeast Ireland*, The Geologists' Association, London, 45 pp.

Clarke, F. W., and Washington, H. S., 1924, The composition of the earth's crust, *U.S. Geol. Survey Prof. Paper*, No. 127, 117 pp.

Cline *et al.*, 1955, Soil Survey of the Territory of Hawaii, *U.S. Dept. Agriculture Soil Series* 1939, No. 25, 644 pp. and separate maps.

Cole, W. F., 1943, X-ray analysis (by the powder method) and microscopic examination of the products of weathering of the Gingin Upper Greensand, *Royal Soc. Western Australia Jour.*, **27**, 229–243.

Connor, Jane, Shimp, N. F., and Tedrow, J. C. F., 1957, A spectrographic study of the distribution of trace elements in some podzolic soils, *Soil Science*, **83**(1), 65–73.

Corbet, S. A., 1933, *Biological Processes in Tropical Soils*, W. Heffer & Sons, Cambridge, 156 pp.

Costin, A. B., Hallsworth, E. G., and Woof, Marion, 1952, Studies in pedogenesis in New South Wales. III, The alpine-humus soils, *Jour. Soil Sci.*, **3,** 190–217.

Dean, L. A., 1957, Plant nutrition and soil fertility, *U.S. Dept. Agriculture Yearbook for 1957, Soils*, pp. 80–85.

Dickson, B. A., and Crocker, R. L., 1953, A chronosequence of soils and vegetation near Mt. Shasta, California. I, Definition of the ecosystem investigated and features of the plant succession: *Jour. Soil Sci.*, **4,** 123–141; II, The development of forest floors and the carbon and nitrogen profiles of the soils, *ibid.*, 142–154; III, Some properties of the mineral soils, *ibid.*, **5,** 173–191.

Doeglas, D. J., 1949, Loess, an aeolian product, *Jour. Sed. Petrology*, **19,** 112–117.

Dorfman, M. D., 1958, Geochemical characteristics of weathering processes in nepheline syenites of Khibina Tundra, *Geochemistry*, No. 5, pp. 537–551.

Drew, J. V., and Tedrow, J. C. F., 1957, Pedology of an arctic brown profile near Point Barrow, Alaska, *Soil Sci. Soc. America Proc.*, **21,** 336–339.

Edelman, C. H., 1950, *Soils of the Netherlands*, North-Holland Publ. Company, Amsterdam, 177 pp.

Ellis, B. S., 1952, Genesis of a tropical red soil, *Jour. Soil Sci.*, **3,** 52–62.

England, C. B., and Perkins, H. F., 1959, Characteristics of three reddish brown lateritic soils of Georgia, *Soil Science*, **88,** 294–302.

Eriksson, E., 1958, The chemical climate and saline soils in the arid zone, *Arid Zone Research*, **10,** 147–180 (Proc. Canberra Conference), UNESCO, New York.

Eriksson, E., 1959, The yearly circulation of chloride and sulfur in nature; meteorological, geochemical, and pedological implications, I, *Tellus*, **12,** 375–403.

Eyles, V. A., 1952, The composition and origin of the Antrim laterites and bauxites, *Geol. Survey of Northern Ireland Memoir*, 90 pp.

Feustel, I. G., and Dutilly, A., 1939, Properties of soils from North American arctic regions, *Soil Science*, **48,** 183–198.

Fleischer, M., 1967, Fluoride content of groundwater in the coterminous United States, *Geol. Soc. America Special Paper*, No. 90, p. 65.

Fosberg, F. R., 1953, Vegetation of central Pacific atolls, a brief summary, *Atoll Research Bull.*, No. 23, pp. 1–26 (Natl. Acad. Sci—Natl. Research Council, Washington, D.C.).

Fosberg, F. R., and Carroll, Dorothy, 1965, Terrestial sediments and soils of the northern Marshall Islands, *Atoll Research Bull.*, No. 113, 156 pp.

Fox, C. S., 1935–36, Buchanan's laterite of Malabar and Kanara, *Records Indian Geol. Survey*, **69,** 389–422.

Fripiat, J. J., and Gastuche, M., 1952, Etude physico-chimique des surfaces des argiles, *Inst. Natl. Etude Agronom. Congo Belge* (*INEAC*), *Serie Scientifique*, No. 54, 60 pp. (Brussels).

Garrels, R. M., 1953, Some thermodynamic relations among the vanadium oxides, and their relation to the oxidation state of the uranium ores of the Colorado Plateau, *Amer. Mineral.*, **38,** 1251–1265.

Garrels, R. M., 1954, Mineral species as functions of pH and oxidation–reduction potentials, with special reference to the zone of oxidation and secondary enrichment of sulphite ore deposits, *Geochim. Cosmochim. Acta*, **5,** 153–168.

Garrels, R. M., 1957, Some free energy values from geologic relations, *Amer. Mineral.*, **42,** 780–791.

Garrels, R. M., and Christ, C. L., 1965, *Solutions, Minerals, and Equilibria*, Harper and Row, New York, 450 pp.

Garrels, R. M., and Dreyer, R. M., 1952, Mechanism of limestone replacement at low temperatures and pressures, *Geol. Soc. America Bull.*, **63**, 325–380.

Garrels, R. M., and Howard, Peter, 1959, Reactions of feldspar and mica with water at low temperatures and pressures, *in: Clays and Clay Minerals*, Natl. Conf. Clays and Clay Minerals, 6th Berkeley, Calif., Proc., Ada Swineford (editor): Pergamon Press, New York, pp. 68–88.

Geiger, Rudolph, 1965, *The Climate near the Ground*, Harvard University Press, Cambridge, Mass., 611 pp.

Glazovskaya, M. A., 1950, The weathering of rocks in the snow belt of central Tanshan (Tein Shan), *Trudy Pochv. Inst. Dokuchaeva Akad. Nauk SSSR*, **34**, 28.

Glazovskaya, M. A., 1958, Weathering and primary soil formation in Antarctica, *Nauch. Dolk. vys. Shkol. geol.-geogr. Nauki*, **1**, 63–76.

Glentworth, Robert, 1944, Studies on the soils developed on basic igneous rocks in central Aberdeenshire, *Royal Soc. Edinburgh Trans.*, **61**, 149–170.

Glinka, K. D., 1935, *The Great Soil Groups of the World and Their Development*, Edward Bros., Ann Arbor, Mich., 150 pp. (trans. from the German edition by C. F. Marbut).

Goldberg, E. D., 1957, Biogeochemistry of trace metals, *Geol. Soc. America Memoir*, No. 67, **1**, 345–357.

Goldberg, E. D., and Arrhenius, G., 1958, Chemistry of Pacific pelagic sediments, *Geochim. Cosmochim. Acta*, **13**, 153–213.

Goldich, S. S., 1938, A study on rock weathering, *Jour. Geology*, **46**, 17–58.

Goldschmidt, V. M., 1937, The principles of distribution of chemical elements in minerals and rocks, *Chem. Soc. London Jour.*, pp. 655–673.

Gordon, M., Jr., Tracey, J. I., and Ellis, M. W., 1958, Geology of the Arkansas bauxite region, *U.S. Geol. Survey Prof. Paper*, No. 299, 268 pp.

Graham, E. R., 1941, Acid clay—an agent in chemical weathering, *Jour. Geology*, **49**, 392–401.

Graham, E. R., 1957, The weathering of some boron-bearing materials, *Soil Sci. Soc. America Proc.*, **21**, 505–507.

Grim, R. E., 1953, *Clay Mineralogy*, McGraw-Hill Book Company, New York, 384 pp.

Hale, G. E., 1957, Some aspects of jointing and decomposition in Tasmanian dolerites, *in: Dolerites, A Symposium*, University of Tasmania, pp. 184–196 (mimeo. report).

Hanlon, F. N., 1944, The bauxites of New South Wales, their distribution, composition, and probable origin, *Royal Soc. New South Wales Jour. and Proc.*, **78**, 94–112.

Hardy, F., and Rodrigues, G., 1939a, Soil genesis from andesite in Grenada, BWI, *Soil Science*, **48**, 361–384.

Hardy, F., and Rodrigues, G., 1939b, The genesis of Davidson clay loam, *Soil Science*, **48**, 483–495.

Harrassowitz, H., 1926, Laterit—Materiel und Versuch erdgeschichtlicher Auswertung, *Fortschr. Geol. u. Paleont.*, **4**(14), 1–314.

Harrison, Sir J. B., 1934, *The Katamorphism of Igneous Rocks under Humid Tropical Conditions*, Imperial Bureau of Soil Science, Harpenden, England, 79 pp.

Harriss, R. C., and Adams, J. A. S., 1966, Geochemical and mineralogical studies on the weathering of granitic rocks, *Amer. Jour. Sci.*, **264**, 146–173.

Haseman, F. J., and Marshall, C. E., 1945, The use of heavy minerals in studies of the origin and development of soils, *Missouri Univ. Agric. Expt. Sta. Research Bull.*, No. 387.

Hathaway, J. C., and Carroll, Dorothy, 1964, Petrography of the insoluble residues, *in* Schlanger, S. O., Petrography of the limestones of Guam, *U.S. Geol. Survey Prof. Paper*, No. 403–D, pp. 37–44.

Hathaway, J. C., and Schlanger, S. O., 1965, Nordstrandite ($Al_2O_3 \cdot 3H_2O$) from Guam, *Amer. Mineral.*, **50**, 1029–1037.

Hay, R. L., 1960, Rate of clay formation and mineral alteration in a 4000-year-old volcanic ash on St. Vincent, BWI, *Amer. Jour. Sci.*, **258**, 354–368.

Hem, J. D., 1959, Study and interpretation of the chemical characteristics of natural waters, *U.S. Geol. Survey Water-Supply Paper*, No. 1473, 269 pp.

Hem, J. D., 1968, Aluminum species in water, *in: Trace Inorganics in Water*, Am. Chem. Soc., pp. 98–114.

Hem, J. D., and Cropper, W. H., 1959, Survey of the ferrous–ferric chemical equilibria and redox potentials, *U.S. Geol. Survey Water-Supply Paper*, No. 1459–A, 31 pp.

Hem, J. D., and Roberson, C. E., 1967, Form and stability of aluminum hydroxide complexes in dilute solution, *U.S. Geol. Survey Water-Supply Paper*, No. 1827–A, 55 pp.

Holtedahl, Hans, 1953, A petrographical and mineralogical study of two high-altitude soils from Trollheimen, Norway, *Norsk Geol. Tidsskr.*, **32**, 191–226.

Holmes, R. S., and Brown, J. C., 1957, Iron and soil fertility, *U.S. Dept. Agriculture Yearbook for 1957, Soils*, pp. 111–115.

Hosking, J. S., 1948, The cation exchange capacity of soils and soil colloids. I. Variation with hydrogen ion concentration; II. The contribution from the sand, silt, and clay fractions and organic matter, *Jour. Council Sci. Indust. Research (Australia)*, **21**, 21–50.

Hosking, J. S., and Greaves, G. A., 1936, A soil survey of an area at Gingin, Western Australia, *Royal Soc. Western Australia Jour.*, **22**, 71–112B.

Hough, G. J., and Byers, H. G., 1937, Chemical and physical studies of certain Hawaiian soil profiles, *U.S. Dept. Agric. Tech. Bull.*, No. 584, 26 pp.

Hough, G. J., Gile, P. L., and Foster, Z. C., 1941, Rock weathering and soil profile development in the Hawaiian Islands, *U.S. Dept. Agric. Tech. Bull.*, No. 752, 43 pp.

Huffmann, Helga, 1954, Mineralogische Untersuchungen an fünf Bodenprofilen über Basalt, Muschelkalk, und Buntsandstein, *Heidelberger Beiträge Mineral. u. Petrograph.*, **4**, 67–88.

Humbert, R. P., and Marshall, C. E., 1943, Mineralogical and chemical studies of soil formation from acid and basic igneous rocks in Missouri, *Missouri Univ. Agric. Expt. Sta. Research Bull.*, No. 359, pp. 5–60.

Hutton, J. T., 1958, The chemistry of rainwater, with particular reference to conditions in southeastern Australia, *in: Climatology and Microclimatology*, Proc. Canberra Symposium, UNESCO, pp. 285–289.

Hutton, J. T., and Stephens, C. G., 1956, The paleopedology of Norfolk Island, *J. Soil Sci.*, **7**, 255–267.

Jacks, G. V., 1934, *Imperial Bureau of Soils Tech. Comm.*, No. 29, p. 9.

Jacks, G. V., 1953, Organic weathering, *Science Progress*, **41**(162), 301–305.

Jackson, M. L., and Sherman, G. D., 1953, Chemical weathering of minerals, *Advances in Agronomy*, Vol. 5, Academic Press, New York, pp. 219–318.

Jeffries, C. D., 1937, The mineralogical composition of the very fine sands of some Pennsylvanian soils, *Soil Science*, **43**, 357–364.

Jenny, Hans, 1941, *Factors of Soil Formation*, McGraw-Hill Book Company, New York, 281 pp.

Jenny, Hans, 1951, Contact phenomena and their significance in plant nutrition, *in: Mineral Nutrition of Plants*, Emil Truog (editor): University of Wisconsin Press, Madison, Wis., pp. 107–132.

Jordan, H. V., and Reisenauer, H. M., 1957, Sulfur and soil fertility, *U.S. Dept. Agriculture Yearbook for 1957, Soils*, pp. 107–111.

Kamoshita, Yutaka, 1958, Soils in Japan, *Natl. Inst. Agric. Sci. Tokyo, Misc. Publ. B*, No. 5, 56 pp.

Keller, W. D., 1954, The energy factor in sedimentation, *Jour. Sed. Petrology*, **24**, 62–68.

Keller, W. D., 1957, *The Principles of Chemical Weathering*, Lucas Bros., Columbia, Mo., 111 pp.

Kellogg, C. E., 1938, Soil and society, *in: 1938 Yearbook of Agriculture "Soils and Men,"* pp. 863–886.

Kelly, W. C., and Zumberge, J. H., 1961, Weathering of a quartz diorite at Marble Point, McMurdo Sound, Antarctica, *Jour. Geology*, **69**, 433–446.

Knoch, K., 1929, Die Klimarfaktoren und Uebersicht der Klimatzonen der Erde, *in: Handbuch der Erde*, Edwin Blanck: Vol. 2, pp. 1–53.

Kononova, M. M., 1961, *Soil Organic Matter. Its Nature, Its Role in Soil Formation and in Soil Fertility* (English translation), Pergamon Press, London, 450 pp.

Köppen, W., 1931, *Grundriss der Klimatologie*, Leipzig, 338 pp.

Köppen, W., and Geiger, R., 1954, *Klima der Erde*, Justes Perthes, Darmstadt.

Krauskopf, K. B., 1967, *Introduction to Geochemistry*, McGraw-Hill Book Company, New York, 721 pp.

Krebs, R. D., and Tedrow, J. C. F., 1957, Genesis of three soils derived from Wisconsin till in New Jersey, *Soil Science*, **83**, 207–218.

Krumbein, W. C., and Pettijohn, F. J., 1938, *Manual of Sedimentary Petrolography*, Appleton-Century-Crofts, New York, 549 pp.

Krumbein, W. C., and Tisdel, F. W., 1940, Size distribution of source rocks of sediments, *Amer. Jour. Sci.*, **238**, 296–305.

Kunze, G. W., and Jeffries, C. D., 1953, X-ray characteristics of clay minerals as related to potassium fixation, *Soil Sci. Soc. America Proc.*, **17**, 242–244.

Kunze, G. W., and Oakes, H., 1957, Field and laboratory studies of the Lufkin soil, a Planosol, *Soil Sci. Soc. Amer. Proc.*, **21**, 330–335.

Le Grand, H. E., 1958, Chemical character of water in igneous and metamorphic rocks of North Carolina, *Econ. Geology*, **53**, 178–189.

Le Riche, H. H., 1959, Molybdenum in the Lower Liassic of England and Wales in relation to the incidence of teart, *Jour. Soil Sci.*, **10**, 133–136.

Leighton, M. M., and Willman, H. B., 1950, Loess formations of the Mississippi Valley, *Jour. Geology*, **58**, 599–623.

Lombaard, B. V., 1934, On the differentiation and relationships of the rocks of the Bushveld Complex, *Geol. Soc. South Africa Trans.*, **37**, 5–52.

Lucas, J., and Ataman, G., 1968, Mineralogical and geochemical study of clay mineral transformations in the sedimentary Triassic Jura Basin (France), *Clays and Clay Minerals*, **16**(5), 365–372.

Magistad, O. C., 1925, The aluminum content of the soil solution and its relation to soil reaction and plant growth, *Soil Science*, **20**, 181–213.

Makeyev, O. V., and Nogina, N. A., 1958, Genus of sod-forest soils on residual alluvial traps, *Soviet Soil Science*, No. 7, pp. 778–786.

Malyuga, D. P., 1964, *Biogeochemical Methods of Prospecting*, Consultants Bureau, New York, 205 pp.

Marbut, C. F., 1927, A scheme for soil classification, *1st Internl. Congress Soil Science Proc.*, Pt. 4, pp. 1–31.

Marshall, C. E., see Haseman, F. J., and Marshall, C. E.

Martell, A. E., 1957, The chemistry of metal chelates in plant nutrition, *Soil Sci.*, **84**, 13–26.

Martin, J. F., and Doyne, H. C., 1927, Laterite and lateritic soils in Sierra Leone, I, *Jour. Agric. Science*, **17**, 530–547; II: *ibid.*, **20**, 135–143.

Mason, B., 1958, An example of climatic control of land utilization, *Arid Zone Research*, **11,** 188–194 (Proc. Canberra Conference), UNESCO, New York.

McCulloch, J. S. G., 1959, Soil temperatures near Nairobi, 1954–1955, *Quart. Jour. Royal Meteorol. Soc.*, **85,** 51–58.

McLaughlin, R. J. W., 1955, Geochemical changes due to weathering under varying climatic conditions, *Geochim. Cosmochim. Acta*, **8,** 109–130.

Merrill, G. P., 1913, *Rocks, Rock Weathering, and Soils:* The Macmillan Co., New York.

Middleton, H. E., 1930, Properties of soils which influence soil erosion, *U.S. Dept. Agric. Tech. Bull.*, No. 178, pp. 1–16.

Millar, C. E., and Turk, L. M., 1951, *Fundamentals of Soil Science*, John Wiley & Sons, New York.

Miller, J. P., 1952, A portion of the system calcium carbonate–carbon dioxide–water, with geological implications, *Amer. Jour. Sci.*, **250,** 161–203.

Miller, J. P., 1961, Solutes in small streams draining single rock types, Sangre de Cristo Range, New Mexico, *U.S. Geol. Survey Water-Supply Paper*, No. 1535–F, 23 pp.

Millot, G., and Bonifas, M., 1955, Transformations isovolumetriques dans les phenomenes de lateritisation et de bauxitisation, *Univ. Strasbourg, Service de la Carte géologique d'Alsace et de Lorraine, Bull.*, **8**(1), 3–20.

Mitchell, R. L., 1964, Trace elements in soils, *in: Chemistry of the Soil*, 2nd ed., Firmin E. Bear (editor): Reinhold Publishing Corp., New York, pp. 320–368.

Mohr, E. J. C., 1944, *The Soils of Equatorial Regions*, J. W. Edwards, Ann Arbor, Mich. (trans. by R. L. Pendleton).

Mohr, E. J. C., and van Baren, F. A., 1954, *Tropical Soils, a Critical Study of Soil Genesis as Related to Climate, Rocks, and Vegetation*, Interscience, New York.

Moorhouse, W. W., 1959, *The Study of Rocks in Thin Section*, Harper and Row, New York, 514 pp.

Morey, G. W., Fournier, R. O., and Rowe, J. J., 1964, The solubility of amorphous silica at 25°, *Jour. Geophys. Research*, **69,** 1995–2002.

Muir, A., Anderson, B., and Stephen, I., 1957, Characteristics of some Tanganyika soils, *Jour. Soil Sci.*, **8,** 1–18.

Mulchay, M. J., and Hingston, F. J., 1961, Soils of the York-Quairading area, Western Australia, in relation to landscape evaluation. *Commonwealth Sci. Indust. Research Organization (Australia) Soil Pub.*, No. 17, 43 pp.

Murata, K. J., 1943, Significance of internal structure in gelatinizing silicate minerals, *U.S. Geol. Survey Bull.*, No. 950, pp. 25–33.

Murata, K. J., 1966, An acid fumarolic gas from Kilauea Iki, Hawaii, *U.S. Geol. Survey Prof. Paper*, No. 537–C, 6 pp.

Nash, V. E., and Marshall, C. E., 1956, The surface reactions of silicate minerals. I, The reaction of feldspar surfaces with acid solutions, *Missouri Univ. Agric. Expt. Sta. Research Bull.*, No. 613, 36 pp.; II, *ibid.*, No. 614, 36 pp.

Nelson, L. A., Kunze, G. W., and Godfrey, C. L., 1960, Chemical and mineralogical properties of San Saba clay, a grumosol, *Soil Science*, **89,** 122–131.

Nye, P. H., 1955, Some soil-forming processes in the humid tropics. IV, The action of the soil fauna, *Jour. Soil Sci.*, **6,** 73–83.

Nye, P. H., 1959, Some effects of natural vegetation on the soils of West Africa and on their development under cultivation, *in: Tropical Soils and Vegetation, Humid Tropics Research*, Proc. Abidjan Symposium, UNESCO, pp. 59–63.

Pannebokke, C. R., 1959, A study of some soils in the dry zone of Ceylon, *Soil Science*, **87,** 67–74.

Patterson, S. H., 1967, Bauxite reserves and potential aluminum resources of the world, *U.S. Geol. Survey Bull.*, No. 1228, 178 pp.

Paver, H., and Marshall, C. E., 1934, The role of aluminum in the reaction of the clays, *Jour. Soc. Chem. Industry (London)*, **53**, 750–760.

Perkins, H. F., and Purvis, E. R., 1954, Soil and plant studies with chelates of ethylenediaminetetra-acetic acid (versene), *Soil Science*, **78**, 325–330.

Pettijohn, F. J., 1957, *Sedimentary Rocks*, 2nd ed., Harper and Row, New York.

Péwé, T. L., 1955, Origin and character of the upland silt near Fairbanks, Alaska, *Geol. Soc. America Bull.*, **66**, 699–724.

Polynov, B., 1937, *The Cycle of Weathering*, Murby, London, 220 pp. (transl. from Russian by A. Muir).

Popov, A. A., 1958, Some problems of chestnut-soil genesis in the semihumid subtropical region of Eastern Caucasus, *Pochvovedenie*, pp. 105–109 [*Soviet Soil Science*, No. 3, pp. 310–313 (English trans.)].

Prescott, J. A., 1931, The soils of Australia in relation to vegetation and climate, *Council Sci. Indust. Research (Australia) Bull.*, No. 52, 82 pp.

Prescott, J. A., 1949, A climatic index for the leaching factor in soil formation, *Jour. Soil Sci.*, **1**, 9–19.

Prescott, J. A., and Pendleton, R. L., 1952, *Laterite and Lateritic Soils*, Commonwealth Agric. Bureau, Harpenden, England, 51 pp.

Preusse, H. H., 1957, Untersuchungen über die Tonkolloide verschiedener hessischer Boden, *Notizbl. hess. L-Amt Boden Forsch.*, **85**, 334–379.

Prider, R. T., 1966, The lateritized surface of Western Australia, *Australian Jour. Sci.*, **28**, 443–451.

Raggatt, H. G., Owen, H. B., and Hills, E. S., 1945, The bauxite deposits of the Boolarra–Mirboo north area, South Gippsland, Victoria, *Commonwealth (Australia) Dept. Supply and Shipping Mineral Resources Bull.*, No. 14, 52 pp.

Reeseman, A. L., Pickett, E. E., and Keller, W. D., 1969, Aluminum ions in aqueous solutions, *Amer. Jour. Sci.*, **267**, 99–113.

Reiche, Parry, 1950, A survey of weathering processes and products, *New Mexico Univ. Publ. in Geology*, No. 3, 95 pp.

Reifenberg, A., 1952, The soils of Syria and the Lebanon, *Jour. Soil Sci.*, **3**, 68–88.

Reifenberg, A., and Ewbank, E. K., 1933, Investigation of soil profiles from Cyprus. I, Profiles of soils over limestone and serpentine, *Empire Jour. Exper. Agric.*, **1**, 85–96; II, Soils over diabase, gabbro-norite, and pillow-lava, *ibid.*, 156–164.

Reuther, W., 1957, Copper and soil fertility, *U.S. Dept. Agric. Yearbook for 1957, Soils*, pp. 128–135.

Rice, T. D., and Alexander, L. T., 1938, The physical nature of soil, *U.S. Dept. Agric. Yearbook for 1938, Soils and men*, pp. 887–910.

Rich, C. I., and Obenshain, S. S., 1955, Chemical and clay mineral properties of a red–yellow podzolic soil derived from muscovite schist, *Soil Sci. Soc. America Proc.*, **19**, 334–339.

Roberts, R. C., 1942, Soil survey of Puerto Rico, *U.S. Dept. Agric. Bur. Plant Industry Soil Series 1936*, No. 8, 503 pp.

Robinson, W. O., Edgington, G., and Byers, H. G., 1935, Chemical studies of infertile soils derived from rocks high in magnesium and generally high in chromium, *U.S. Dept. Agric. Tech. Bull.*, No. 471, 28 pp.

Rodrigues, G., 1954, Fixed ammonia in tropical soils, *Jour. Soil Sci.*, **5**, 264–274.

Rodrigues, G., and Hardy, F., 1947, Soil genesis from a sedimentary clay in Trinidad, *Soil Science*, **64**, 127–142.

Rolfe, B. N., and Jeffries, C. D., 1952, A new criterion for weathering in soils, *Science*, **116,** 599–600.

Rudolph, E. D., 1966, Terrestial vegetation of Antarctica: past and present studies, *Antarctic Research Series*, No. 8, pp. 109–124.

Russel, D. A., 1957, Boron and soil fertility, *U.S. Dept. Agric. Yearbook for 1957, Soils*, pp. 121–128.

Ruxton, B. P., and Berry, L., 1957, Weathering of granite and associated erosional features in Hong Kong, *Geol. Soc. America Bull.*, **68,** 1263–1292.

Sakamoto, Takeo, 1960, Ecological studies in rain-forest in northern Suriname, *Verh. Konin. Netherlands Akad. van Wetenschappen, Afd. Natuurkinde, Tweede Rekke*, **53**(1), 1–11.

Schatz, A., Cheronis, N. D., Schatz, V., and Trelawney, G. J., 1954, Chelation (sequestration) as a biological weathering factor in pedogenesis, *Pennsylvania Acad. Sci. Proc.*, **28,** 44–57.

Seatz, L. F., and Jurinak, J. J., 1957, Zinc and soil fertility, *U.S. Dept. Agric. Yearbook for 1957, Soils*, pp. 115–121.

Seelye, F. T., Grange, L. I., and Davis, L. H., 1938, The laterites of western Samoa, *Soil Science*, **46,** 23–31.

Shacklette, H. T., and Cuthbert, M. E., 1967, Iodine content of plant groups as influenced by variation in rock and soil type, *Geol. Soc. America Special Paper*, No. 90, pp. 31–46.

Sherman, G. D., 1955, Chemical and physical properties of Hawaiian soils, *in: Soil Survey, Territory of Hawaii, U.S. Dept. Agric. Soil Survey Series 1939*, No. 25, pp. 110–124.

Sherman, G. D., 1957, Manganese and soil fertility, *U.S. Dept. Agric. Yearbook for 1957, Soils*, pp. 135–139.

Sherman, G. D., and Uehara, G., 1956, The weathering of olivine basalt in Hawaii and its pedogenic significance, *Soil Sci. Soc. America Proc.*, **20,** 337–340.

Short, N. M., 1961, Geochemical variations in four residual soils, *Jour. Geology*, **69**(5), 534–571.

Shul'gin, A. M., 1957, *The Temperature Regime of Soils*, Israel Program for Scientific Translations, Ltd., Jerusalem, 1965.

Siever, R., 1962, Silica solubility, 0°–200°C, and the diagenesis of siliceous sediments, *Jour. Geology*, **70,** 127–150.

Simpson, E. S., 1912, Notes on laterite in Western Australia, *Geol. Magazine*, **9**(n.s. decade V), 399–406.

Smith, G. D., 1942, Illinois loess—variations in its properties and distribution, *Univ. Illinois Agric. Expt. Sta. Bull.*, No. 490.

Smith, J., 1956, Some moving soils in Spitzbergen, *Jour. Soil Sci.*, **7,** 10–21.

Starkey, R. L., and Wight, K. M., 1945, *Anaerobic Corrosion of Iron in the Soil*, Final Rept. Amer. Gas Assoc. Iron Corrosion Research Fellowship, Amer. Gas Assoc., New York, 108 pp.

Stephen, Ian, 1952, A study of rock weathering with reference to the soils of the Malvern Hills. I, Weathering of biotite and granite, *Jour. Soil Sci.*, **3,** 20–33; II, Weathering of appenite and "ivy-scar rock," *ibid.*, **3,** 219–237.

Stephen, Ian, 1953, A petrographic study of a tropical black earth and grey earth from the Gold Coast (Ghana), *Jour. Soil Sci.*, **4,** 211–219.

Stephens, C. G., and Hutton, J. T., 1954, A soil and land-use study of the Australian Territory of Norfolk Island, South Pacific Ocean, *Commonwealth Sci. Indust. Research Organization (Australia) Soils and Land-Use Series*, No. 12, 28 pp.

Stout, P. R., and Johnson, C. M., 1957, Trace elements, *U.S Dept. Agric. Yearbook for 1957, Soils,* pp. 139–150.

Suwa, Kanenori, Matsuzawa, Isao, IIda, Chuzo, and Yamasaki, Kazuo, 1958, Mineralogical and geochemical weathering of a quartz porphyry: *Nagoya Univ. Jour. Earth Science,* **6,** 75–100.

Swaine, D. J., 1955, The trace element content of soils, *Commonwealth Bureau Soil Science Tech. Commun.,* No. 48, 157 pp.

Swindale, L. D., and Jackson, M. L., 1956, Genetic processes in some residual podzolized soils of New Zealand, *VI Congr. int. Sol. Rapp. E.* 233–239.

Swineford, Ada, and Frye, J. C., 1946, Petrographic comparison of Pliocene and Pleistocene volcanic ash from western Kansas, *State Geol. Survey Kansas Bull.,* No. 64, Pt. 1, pp. 1–32.

Swineford, Ada, and Frye, J. C., 1955, Petrographic comparison of some loess samples from western Europe with Kansas loess, *Jour. Sed. Petrology,* **25,** 3–23.

Tamhane, R. V., and Namjoshi, N. G., 1959, A comparative study of black soils formed from different parent materials, *Jour. Indian Soc. Soil Science,* **7,** 49–63.

Teakle, L. J. H., 1937, The red and brown hard-pan soils of the Acacia semidesert scrub of Western Australia, *Jour. Dept. Agric. Western Australia,* **13**(second series)(4), 480–493.

Tedrow, J. C. F., and Ugolini, F. C., 1966, Antarctic soils, *Antarctic Research Series,* **8,** 161–177.

Thornthwaite, C. W., 1958, Introduction to arid zone climatology, *Arid Zone Research,* UNESCO, **11,** 15–22.

Thorp, J., Johnson, W. M., and Reed, E. C., 1951, Some post-Pliocene buried soils of central United States, *Jour. Soil Sci.,* **2,** 1–19.

Tiller, K. G., (1957), The geochemistry of basaltic tuff and basalt and associated soils in the Mt. Burr area of southeastern South Australia, *Commonwealth Sci. Indust. Research Organization (Australia), Div. Soils, Div. Rept.,* March 1957, 41 pp. (mimeo).

Tiller, K. G., 1958, Geochemistry of some basaltic materials-associated soils of southeastern South Australia, *Jour. Soil Sci.,* **9,** 225–241.

Tiller, K. G., 1959, Distribution of some trace elements in the soils developed on dolerite in Tasmania, *Commonwealth Sci. Indust. Research Organization (Australia), Div. Soils, Div. Rept.,* June 1958, 32 pp. (mimeo).

Toth, S. J., 1964, The physical chemistry of soils, *in: Chemistry of the Soil,* 2nd ed., Firmin E. Bear (editor): Reinhold Publishing Corp., New York, pp. 142–205.

Tourtelot, H. A., 1962, Preliminary investigation of the geologic setting and chemical composition of the Pierre shale, Great Plains region, *U.S. Geol. Survey Prof. Paper,* No. 390, 74 pp.

Trewartha, G. T., 1954, *An Introduction to Climate,* 3rd ed. (Plate II), McGraw-Hill Book Company, New York.

Tyturin, I. V., 1949, The geographical laws of humus formation, *in: Proceedings of the Jubilee Session Commemorating the Centenary of V. V. Dokuchaev's Birth,* Izd. Akad. Nauk SSSR (in Russian).

Underwood, E. J., 1962, *Trace Elements in Animal and Human Nutrition,* 2nd ed.: Academic Press, New York, 429 pp.

Underwood, E. J., and Filmer, J. F., 1935, Enzootic marasmus. The determination of the biologically potent element (cobalt) in limonite, *Austr. Vet. Jour.,* **11,** 84–92.

U.S. Department of Agriculture, 1938, Soils of the United States, *in: Soils and Men, Yearbook of 1938,* pp. 1019–1161.

U.S. Department of Agriculture, 1951, *Soil Survey Manual*, Handbook No. 18, 502 pp.

U.S. Department of Agriculture, 1957, *Soil*, Yearbook for 1957, 784 pp.

U.S. Department of Agriculture, 1960, *Soil Classification, a Comprehensive System*, 7th approximation, 265 pp.

U.S. Salinity Laboratory Staff, 1954, Diagnosis and improvement of saline and alkali soils, *U.S. Dept. Agriculture Handbook* No. 60, 160 pp.

Vageler, P., 1913, Physikalische und chemische Vorgänge bei der Bodenbildung in den Tropen, *Mitt. dtsch, Landw. Ges.*, **26**, 396.

Válek, Bohumil, 1959, Die Boden des Süd-Kaldidalur, Gebiet auf südwest Island und ihre Vegetationbeziehungen, *Rozpravy Ceskoslovenske Akademie, Ved.*, **8**(69), 3–34.

van Baren, J., 1931, Properties and constitution of a volcanic soil built in 50 years in the East Indian Archipelago, *Meded. Landbouw Hogesch. Wageningen*, No. 34, 29 pp.

van Doormaal, J. C. A., 1945, *Onderzoekingen betreffende de Lossgrunden van Zuid-Limburg*, J. H. Gottmer, Haarlem, Netherlands, 94 pp.

Vinogradov, A. P., 1959, *The Geochemistry of Rare and Dispersed Chemical Elements in Soils*, Consultants Bureau, New York, 209 pp. (2nd revised ed., English trans.).

Vinogradov, A. P., 1964, Provinces biogeochimique et leur role dans l'evolution organique, *in: Advances in Organic Geochemistry, Earth Science Series*, No. 15, Pergamon Press, New York, pp. 317–337.

Waegemans, G., 1954, Les laterites de Gimbi (Lower Congo), *Pub. inst. natl. etude agron. Congo Belge (INEAC), serie scientifique*, No. 60, 27 pp.

Waksman, S. A., 1938, *Humus: Origin, Chemical Composition and Importance in Nature*, 2nd ed., Williams and Wilkins Co., Baltimore, Md.

Wall, J. R. D., *et al.*, 1962, Nordstrandite in soil from Sarawak, Borneo, *Nature*, **196**, 264–265.

Walther, Johannes, 1915, Laterit in Westaustralien, *Z. dtsch. Geolog. Ges. B. Monatsberichte*, **67**(4), 113–140.

Wentworth, C. K., 1922. A scale of grade and class terms for clastic sediments, *Jour. Geology*, **27**, 507–522.

Whiteside, E. A., and Marshall, C. E., 1944, Mineralogical and chemical studies of the Putnam silt loam, *Missouri Univ. Agric. Expt. Sta. Research Bull.*, No. 386.

Williams, H., Turner, F. J., and Gilbert, M. C., 1958, *Petrography: an Introduction to the Study of Rocks in Thin Sections*, W. H. Freeman & Company, San Francisco, 406 pp.

Wurman, E., 1960, A mineralogical study of a gray-brown podzolic soil in Wisconsin derived from a glauconitic sandstone, *Soil Science*, **89**, 38–44.

Yamasaki, K., Iida, C., and Yokoi, H., 1955, A spectrographic determination of the distribution of trace elements in a granodiorite and in its weathering products, *Nagoya Univ. Jour. Earth Science*, **3**(1), 58–64.

Zaboyeva, I. V., 1958, Gley-podzolic soils, *Soviet Soil Science*, No. 3, pp. 237–244.

Zeller, D. E. (editor), 1968, The stratigraphic succession in kansas, *State Geol. Survey Kansas Bull.*, No. 189, 81 pp.

Zeuner, F. E., 1949, Frost soils on Mount Kenya and the relation of frost soils to aeolian deposits, *Jour. Soil Sci.*, **1**, 20–30.

INDEX

Abrasion pH, 96
Acid attack, 101
Age relationships, 139, 140
Albedo effect, 131
Algae, 128
Alluvial soils, 67
  geographic distribution, 67, 68
Alteration, 5, 175
Alumina gel, 106
Alumina solubility, 100
Aluminosilicates, 97
Amount of chemical weathering, calculating, 69
Animal nutrition, 161
Azonal soils, 39
Barth's calculations, 69
Base saturation, 110
Bauxite, 58
Bentonite beds, 61
Billy, 56
Biogeochemical prospecting, 153
Biogeochemical provinces, 151–153
Biological activity, 117
Bog soils, 53
Boron in plant nutrition, 158
Brown soils, 50, 51
Calcification, 40
Calculations of chemical weathering amounts, 69
Carbonation, 101
Carbonic acid formation, 101
Chelation, 124
Chemical analysis of weathering, 70
Chemical characteristics of elements, 148
Chemical climate, 18
Chemical weathering, 19, 89
Chernozem soils, 50
Chestnut soil, 50
Chevulation, 126
Chlorine in plant nutrition, 160
Clay development, 75
  chemical analysis, 79
  index-mineral calculations, 76
Clay minerals, 35–38, 107
Climate, 9
  Köppen classification, 9–11
  symbols, 11
Climatic index, P/E^m, 9
Climatic zones
  main chemical process, 11
  weathering characteristics, 12, 13
Cobalt in animal nutrition, 162
Copper
  in animal nutrition, 163
  in plant nutrition, 157
Deep weathering, 20
Desert soils, 51
Disequilibrium and equilibrium of igneous rocks, 5
Earth's crust (*see* lithosphere)
Ecology, 121
Element mobility, 103
Erosion, 67
Erosion remnants, 144
Ethylenediaminetetra-acetic acid (EDTA), 127
Feldspars in hydrolysis, 104, 105
Five variables of weathering, 15
Fluorine
  in animal nutrition, 165
  in human nutrition, 166
Fossil soils, 140, 141
Fulvic acids, 125
Geochemical weathering
  effectiveness, 21
  general, 19, 20
  process, 23
Glaciation, 136
Gleization, 40
Grain packing, 86
Grain size, 85
  in soils, 87
Great soil groups, 37, 41, 42
Groundwater
  fluorine content, 168
  pH, 97

Groundwater podzols, 52
Half-bog soils, 52
Humic acids, 125
Humus
  development, 122
  in soils, 123
Hydration, 106
Hydrolysis, 103
  of basic rocks, 105
Igneous minerals, variation, 57
Index mineral, 69, 75
Insoluble material, 82
Intrazonal soils, 52
Iodine
  in animal nutrition, 165
  in human nutrition, 166, 167
Ion exchange, 107, 110
Ion-exchange capacity, 109, 112
Ion fixation, 111
Ionic potentials, 148
Iron
  in animal nutrition, 162
  in plant nutrition, 156
Isomorphic replacement, 146
Jenny's five variables of weathering, 15
Jointing, 81
Lahars, 59
Laterite, 54
  chemical composition, 57, 58
  development, 55, 56
  profile, 54
  mineral composition, 57
Lateritic soils, 53
Lateritization, 40, 53
Lithosols, 53
Lithosphere composition
  accessory minerals, 3
  chemical elements, 2
  feldspars, 2
  igneous rocks, 2
  metamorphic rocks, 4, 32
  rock-forming minerals, 3
  sedimentary rocks, 4, 31, 32
Lithosphere, description, 1
Leaching, 6, 8, 96, 102
  types of water, 94
Leaching factor, *P/sd*, 9
Loess, 61, 144
  "cold loess," 64
  grain-size distribution, 63
  "hot loess," 64
  Mississippi River basin, 62
  Netherlands, 64
Macrofauna, 120
Macroflora, 119
Manganese in plant nutrition, 157
Marbut's soil classification, 41, 42
Mature soil, 26
Mechanical weathering, 81
  in Antarctica, 83
Microclimate, 11, 14, 15
Microfauna, 120
Microflora, 119
Micronutrient deficiencies in humans, 166
Microorganism abundance, 120
Mineral alteration, 170
Mineral bonds, 90
Mineral clogging, 111
Mineral stability, 101, 169
Mineral transformations, 169
Mineralogical analysis of weathering, 72
Molybdenum
  in animal nutrition, 162
  in plant nutrition, 159
Monalto, 72
Mosses, 127
Munsell Soil Color Chart, 29
Noncalcic brown soils (*see* shantung soils)
Organic matter, 117
Organic weathering, 127
Oxidation and reduction, 112
Oxidation potential, 112
Parna, 64
Peaty soils, 40
Pedalfers, 42, 102
Pedocals, 42, 102
Pedochemical weathering, 21
  intensity, 23
  process, 23
Pedon, 25
Petrographic provinces, 150, 151
Physical weathering, 19 , 81
Planosols, 51, 52
Plant nutrition, 155
Podzolic soils
  brown, 47
  gray-brown, 49
  red-yellow, 49, 50
Podzolization, 40, 53
Podzols, 47
Pore space in soils, 88
Prairie soils, 49
Radiant energies, 131
Rainwater
  composition, 94
  pH, 94
Rates of soil formation, 135
Red desert soils, 51
Reddish brown soils, 51
Reduced soils, 114
Reduction, 114
Regional soil patterns, 46

Rendzina, 52
Rock disintegration, 81
  grain-size distribution, 82
Salinization, 40
Sand, 65
  grain-size analysis, 66
  size, 65
Saprolite, 19, 20
  developed from gneiss, 20
  developed from schist, 20
  development, 171
Saprolitization, Western Australia, 21
Selenium in animal nutrition, 163, 164
Selenium toxicity, 164
Shallow soils, 53
Shantung soils, 51
Sheet erosion, 67
Sierozem soils, 51
Silica gel, 101
Silica solubility, 100, 170
Sodium in plant nutrition, 161
Soil, definition, 25
Soil ages, 135, 136
Soil biota, 119
Soil classification, 26, 41
Soil colors, 29
Soil descriptions, 43
Soil formation, 19
  high rainfall areas, 16
  low rainfall areas, 16
  temperature, 16
Soil horizons
  A, B, C, D, names, 26, 39
  general, 26
Soil maps, 43
Soil profile, 26, 39
  nomenclature, 26–28
Soils (*see* individual soils)
Solar radiation, 132
Solonchak soils, 52
Solonetz soils, 52
Stability series, 148, 149
Steady-state chemical environment, 6
Steady state
  in soil formation, 15
  in weathering, 16
Sulfur in plant nutrition, 159
Temperature variation, 129
Terra rosa, 59
Thermal regimes, 132
Trace element
  amounts, 145
  behavior in weathering, 148
  contents, 149, 150
  distribution, 147
  sources, 145
  relationship to plant and animal nutrition, 154, 155
  relationship to human nutrition, 155, 166
Tropical soil fauna, 121
U. S. Department of Agriculture soil classification, revised, 44, 46
  to 1960, 44, 45
Volcanic ash, 59
  chemical composition, 60
Weathering
  biological activity, 21
  climatic conditions, 8, 9
  Eh values, 7
  environment, 7
  fundamental process, 7, 8
  granite, 23, 24
  mineral products, 30, 31
  pH, 7
  principal products, 8
  residual minerals, 33, 34
  variables, 15
Weathering by water, 89
  groundwater, 93
  rainwater, 94
  (*see also* leaching)
Weathering intensity, 149
Weathering potential index, WPI, 34, 35
Weathering sequence, 172
Weathering temperatures, 130
Wentworth scale, 65
Wiesenboden, 52
X-ray diffraction
Zinc
  in animal nutrition, 162
  in plant nutrition, 157
Zonal soils, 39